Also by Roger Ellman:

_____   _____

* THE ORIGIN AND ITS MEANING

ON THE ORIGIN OF THE UNIVERSE AND ITS MECHANICS,
THE MECHANISM AND ORIGINOF INTELLIGENCE,
AND THE IMPLICATIONS FOR THE INDIVIDUAL AND SOCIETY

_____   _____

* ON THE NATURE OF MATTER

THE ORIGIN OF THE UNIVERSE CREATED MATTER
FUNDAMENTALLY WAVE IN NATURE, NOT PARTICULATE

_____   _____

* THE PHILOSOPHIC PRINCIPLES OF RATIONAL BEING

ANALYSIS AND UNDERSTANDING OF
REALITY, TRUTH, GOODNESS, JUSTICE, VIRTUE, BEAUTY,
HAPPINESS, LOVE, HUMAN NATURE, SOCIETY, GOVERNMENT,
EDUCATION, DETERMINISM,
FREE WILL, AND DEATH

_____   _____

* GRAVITICS

THE PHYSICS OF THE BEHAVIOR AND CONTROL OF GRAVITATION

_____   _____

* THE TROUBLE WITH THE HUBBLE LAW

AN ALTERNATIVE TREATMENT OF REDSHIFTS IS EVIDENCED BY
FOUR INDEPENDENT OBSERVED COSMOLOGICAL EFFECTS.

* RESOLUTION OF THE "SPOOKY" PROBLEMS OF QUANTUM MECHANICS

THE WAVE NATURE OF PARTICLES AND LIGHT RESOLVES EINSTEIN'S "SPOOKY" ENTANGLEMENT PROBLEM AND THE "WEIRDNESS" OF QUANTUM MECHANICS

———————————   ———————————

Earth   Proxima b

$$a_{Grav} = G \cdot M / d2$$

HOW TO TRAVEL TO AND EXPLORE MARS OR PROXIMA CENTAURI - B

Use of fuel is impractical for space exploration. The only fuel-free acceleration for space exploration is by controlled gravitation, and that is the only means for a flying vehicle to explore a distant planet and the only means of travel to it.

Such control of gravitation is the subject and purpose of this book.

Just as the sail-driven ships of past centuries explored the world with fuel-free travel by controlled use of the wind; a new gravitation technology will enable fuel-free exploration of space by control of the ubiquitous gravitational field.

ROGER ELLMAN

Cataloging Data

Ellman, Roger (1932-   )

How To Travel To and Explore Mars or Proxima Centauri b

The physics and cosmological problems and their solution.

Library of Congress Control Number:2019911434

Published by: The-Origin Foundation, Inc.,
                 1401 Fountaingrove Pkwy.
                 Santa Rosa, CA 95403, USA

                 707-537-0257

                 http://www.The-Origin.org

ISBN: 9781088501719

CONTENTS

The-Origin Foundation, Inc. is a non-profit organization founded to foster independent scientific, mathematical, and philosophical research.

The author of the present work, Roger Ellman is the General Director of the foundation.

Roger Ellman has published over fifty professional papers on topics ranging from physics, cosmology, and astrophysics to artificial intelligence and mathematics.

He has presented some of his papers to conferences of / at:

> The American Physical Society [APS], .
> The American Society for the Advancement of Science,
> Cambridge University, United Kingdom
> The Library of Alexandria, Egypt
> The Russian Academy of Natural Sciences, St Petersburg
> The Hungarian Academy of Sciences, Budapest
> A Science Conference in Shang Hai, China

He is author of six books in addition to the present "How to Travel to and Explore Mars or Alpha Centauri".

His education includes graduate studies at Stanford University after graduating from West Point, the United States Military Academy.

The problem of space exploration is the immense distances, so immense that they are measured in the distance that light travels at its great speed for a year, the light-year. Neglecting our "near" moon, whether our space exploration is of Mars or of Alpha Centauri's planet Proxima Centauri b [after Mars the next nearest human habitable planet] the round trip will take at least a year [Mars] or over 10 years [Proxima Centauri b].

That means that the vehicle must support the crew with food, water, elimination of waste, breathable air, medical services, electric power, *etc.* without any support from Earth for a year or more, which means that the vehicle must be massive, and that is just to get there and return home.

The purpose in traveling to Mars or Proxima Centauri b is to explore and scientifically investigate and that requires, in addition to simply getting there, a significant amount of scientific and laboratory equipment. Not only that, a flying vehicle to travel all over the destination planet is essential, which means that the space ship vehicle carrying all of that must be even more massive.

That massive vehicle ship must be accelerated and decelerated throughout the trip without use of fuel. Rockets consume far too much fuel for long distance or long term use and solar sails are far too inflexible and inefficient. The only fuel-free acceleration for space exploration is by controlled gravitation; and that is the only means of a flying vehicle to explore all over the destination planet.

Such control of gravitation is the subject and purpose of this book.

Just as the sail-driven ships of past centuries
explored the world with fuel-free travel
by controlled use of the wind;
a new gravitation technology
enables fuel-free exploration of space
by control of the ubiquitous gravitational field.

Use of fuel is impractical for space exploration propulsion. Fuel is by far the largest, most significant, weight component of rocket propulsion systems. The only fuel-free acceleration propulsion method for space exploration is controlled gravitation. That is the only means for a flying vehicle to explore a distant planet upon arrival there and the only means of getting to the distant planet so as to explore it.

<u>The only solution to the problem of space travel propulsion is Control of Gravitation.</u>

Such Control of Gravitation is the subject and purpose of this book.

THE PROBLEM

Space Exploration Requires Control of Gravitation

> The primary problem of space exploration, whether of Mars or
> Proxima Centauri, b is that
> the distances are so immense.

They are so immense that they are measured in terms of the distance that light travels at its great speed, the greatest speed possible, for one full Earth year. It is for that reason that we are interested in Alpha Centauri. It is the closest star system to us at 4.37 light-years from our Sun; its distance is the least overwhelming of the myriad other cosmic distances. In addition it has the closest ever detected planetary system.

Alpha Centauri is named for the constellation of stars that is named for the Centaur of Greek mythology [upper body of a man with lower body of a horse], and Alpha for the brightest point in that constellation. It consists of three stars: Alpha Centauri A, B, and C. Alpha Centauri A and B are stars similar to our Sun in a binary pair rotating about a common center and for them we have no detected planets in the habitable zone.

Though not visible to the naked eye, Alpha Centauri C [also called Proxima Centauri], a small faint red dwarf about 0.2 light-years from the A-B binary pair is the closest star to our Sun at a distance of 4.24 light-years.

Only one planet has been confirmed for Proxima Centauri, Proxima Centauri b. That planet is slightly larger than the Earth and orbits around Proxima Centauri in the habitable zone.

And, perhaps, that planet's resemblance to our Earth is why we are interested in How to Travel to Alpha Centauri

PROBLEMS SUMMARY

For travel to Alpha Centauri there are two major problems:

[1] Because of the distance it will take over 4 years to get there. The vehicle must support the crew with food, water, elimination of waste, breathable air, medical services, electric power, *etc.* without any support from Earth for at least 9+ years for the round trip. Thus the vehicle must be massive.

1

[2] That massive ship must be accelerated and decelerated throughout the trip without use of fuel. Rockets consume far too much fuel and solar sails are far too inflexible and inefficient. The only fuel-free acceleration is by gravitation.

1

Just as the sail-driven ships of past centuries explored the world with fuel-free travel by controlled use of wind, a new gravitation technology enables fuel-free exploration of space by control of the ubiquitous gravitational field.

THE PROBLEM OF UNDERSTANDING GRAVITATION

Contemporary science has completely failed to comprehend the cause of gravitation, its mechanism, how it works. Einstein became the "hero" of gravitation because of his General Theory of Relativity and the success that it has had in describing and predicting the behavior of gravitation and the various effects that gravitation produces.

But, none of that addresses the mechanism causing the action of gravitation. The closest that Einstein's General Relativity comes to presenting a mechanism of gravitation is its contention that gravitational mass bends or distorts space and that the resulting non-linearity produces the gravitational effects.

However, Einstein's General Relativity offers no explanation of how, why, by what mechanism gravitational mass bends space; nor does it offer any explanation of what space is and how it is susceptible to bending.

The famous Eddington experiment in which the deflection of light by the gravitational field of the Sun was successfully measured was taken as comprehensive proof of Einstein's General Relativity even though it included no understanding of the cause, the mechanism of gravitation.

And, the failure to even have an interest in investigating further into gravitation to seek to achieve an understanding not merely of what gravitation does but, significantly further, how and why it does what it does remains to the present day as a giant impediment to coming to understand gravitation and to turning it to useful application for mankind.

THE TASK AHEAD

In view of the foregoing, that is:

- The necessary massiveness of a vehicle for humans to travel deep space distances requires a controlled gravitational acceleration propulsion system,

and

- The absence of scientific research into the nature, mechanism, and cause of gravitation, let alone research into means of controlling gravitation,

combined leave open the way for development of control of gravitation based upon scientifically developed understanding of how and why gravitation operates.

That develops as follows.

2

Experience shows that everything has a cause and that those causes are themselves the results of precedent causes, *ad infinitum*. Defining and comprehending the causality or mechanism operating to produce an observed behavior is essential to understanding or explaining it.

The comprehensive explanation of the cause and mechanism of gravitation as derived from the origin of the universe is the Modern Newtonian Model of Gravitation. Its development consists of the following steps all thoroughly treated in the following sections.

First Steps – The Nature of Gravitation

The development of the Modern Newtonian Model of Gravitation consists of the following steps. Each step results in new "hard" facts generated directly from prior "hard" facts. The development does not contain nor rely on opinions. Consequently, while it is deemed a "model" it is an exact factual description of what it treats.

1 – How the universe's particles of matter came into existence.

2 – How they came to be propagating an oscillatory outward flow.

3 – The reservoir supply for the substance of the outward flow.

4 – The speed of the outward flow.

5 – A particle's flow encountering another particle slows its outward flow.

6 – The outward flow has momentum.

7 – Gravitation is the momentum reaction to outward flow slowing.

Then, given the nature and mechanism of gravitation, the development of the management and control of gravitation and the development of specific applications of that control proceeds as follows.

Second Steps – The Control of Gravitation

1 – Deflection of light – slowing of its flow.

2 – Light and gravitation are different aspects of the same matter particle flow.

3 – Slit diffraction of light's flow is diffraction of gravitation's flow.

4 – A gravitation deflector.

5 – The amount of the deflection.

6 – The mechanism of gravitational repulsion, anti-gravity.

7 – Gravitation deflection spacecraft and flying vehicles.

> Next: the development of the nature of gravitation from
> how the universe came into be

Physics scientists have yet to offer explanation of the cause of the "Big Bang" – how and why the universe first came into existence.

But, the details and nature of that first event dictated the nature of the universe to which it gave birth.

And, it dictated the nature of Gravitation.

The following presents the details of the source of the "Big Bang" and then forward to the details of the mechanism of gravitation.

THE NATURE OF GRAVITATION

The Origin of Matter is the Origin of Gravitation

> In order to correctly understand the nature of matter it is necessary to consider all of the applicable sources of information and data.
> There are two such sources:
> - The behavior of matter in its various encountered circumstances.
> - The origin of matter – how and from what it came to be.

INTRODUCTION

Until the present the science community has addressed only the first of those two with regard to the nature of matter and the omission of the second has resulted in a major error in the understanding of the nature of matter – the incorrect solution to the problem of the wave nature of matter versus its particle nature.

Causality or mechanism is apparent from observation and experience which show that every thing and every event has a cause, and that those causes are themselves the results of precedent causes, and *ad infinitum*. Defining and comprehending the causality or mechanism operating to produce any contended or proposed scientific truth is essential to authenticating or validating that truth.

HOW THE UNIVERSE'S MATTER CAME TO BE

We are confronted with an apparently insuperable problem. Before the universe there was nothing, absolute nothing. That is the starting point because it naturally occurs; it is the only starting point that requires no cause, no explanation nor justification for its existence. But, that starting point has two impediments to the universe, or anything, coming into existence from it. First is the problem of change from nothing to something without, at least initially, an infinite rate of change, which is impossible. Second is the problem of change from nothing to something without violating conservation, which must be maintained.

The analysis would appear to end at that point, end with the declaration that obviously there cannot be a universe and there is no universe. Except, of course, that we and the universe we inhabit clearly exist at least enough for us to investigate it. Therefore, a solution to the insuperable problem exists. That solution is as follows.

1 - *THE PROBLEM OF INFINITE RATE OF CHANGE*

To avoid a material infinity the rate of change at the moment of the change must have been finite. Rather than an instantaneous jump from nothing to something, no matter how small or "negligible" that something might have been, there had to be a gradual transition at a finite rate of change. Further, the rate of change of that rate of change, the change's second derivative, at that moment had to have been finite, and so on *ad infinitum* for all of the further derivatives.

That requirement means that the form of the change had to have been either a natural exponential or some form of sinusoid. That develops as follows, in which the sought form of the change will be the function $U(t)$ [the "U" for universe, of course].

To illustrate the problem consider the function

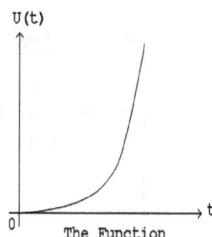

$$(2\text{-}1) \quad \begin{array}{ll} U(t) = 0 & t < 0 \\ U(t) = t^2 & t = 0 \text{ and } t > 0 \end{array}$$

as a theoretical candidate for $U(t)$ at the beginning of the universe, which function is graphically depicted at the right.

Its first derivative, also depicted graphically to the right, is

$$(2\text{-}2) \quad \frac{dU(t)}{dt} = 0 \qquad t < 0$$

$$\frac{dU(t)}{dt} = 2 \cdot t \qquad t > 0$$

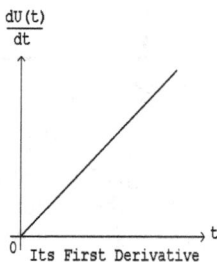

and is unstated for $t=0$ because $dU(t)/dt$ is not smooth there even though $U(t)$ "looks" smooth there.

Figure 2-1a

Now, the second derivative depicted graphically to the right

$$(2\text{-}3) \quad \frac{d^2U(t)}{dt^2} = 0 \qquad t < 0$$

$$\frac{d^2U(t)}{dt^2} = 2 \qquad t > 0$$

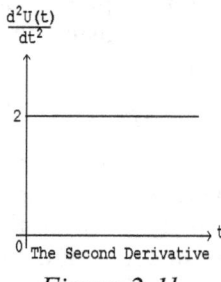

is clearly discontinuous at $t=0$, the instant of the beginning of the universe, where it instantaneously jumps from 0 to 2 as depicted to the right.

Figure 2-1b

The third derivative, which is the rate of change of the second derivative must be infinite at $t=0$ to produce the instantaneous jump from 0 to 2. Clearly, that cannot have happened in the real universe. It is such a condition which is unacceptable in a candidate function for $U(t)$ at the beginning of the universe.

The only way to avoid that condition of an infinite derivative somewhere along the line of successive further derivatives is to have a function with an endless family of finite, non-zero derivatives; that is, some derivatives may be zero at $t=0$ but there must always be further non-zero higher derivatives, which requires that the functional form of every derivative must be non-zero.

One can conceive theoretically of the idea of a function for which all derivatives are non-zero and no two are alike (in a general sense analogous to the pattern of digits in an irrational number), but it is not likely that such a function can exist. In any case the more certain and more simple way to achieve all non-zero derivatives is a repeating derivative function, the two simplest examples of which are as below.

(2-4)
$$\frac{dU(t)}{dt} = \pm U(t) \quad \text{[First derivative = the original function]}$$

(2-5)
$$\frac{d^2U(t)}{dt^2} = \pm U(t) \quad \text{[Second derivative = the original function]}$$

a. *Analysis of Repeating Derivative Functions*

Case (a): Functions Satisfying Equation 2-4

The function meeting this requirement is the natural exponential, ε^t.

(2-6)
$$\varepsilon^t = 1 + t + \frac{t^2}{2!} + \frac{t^3}{3!} + \cdots$$

Taking the first derivative

(2-7)
$$\frac{d[\varepsilon^t]}{dt} = 0 + 1 + \frac{2t}{2!} + \frac{3t^2}{3!} + \cdots$$
$$= 1 + t + \frac{t^2}{2!} + \frac{t^3}{3!} + \cdots = \varepsilon^t$$

so that the original function results as is required by equation *2-4*.

That is the prime case of a function that satisfies the requirement of all derivatives existing in functional form. In general those of this case are as equation *2-8*.

(2-8) $U(t) = A \cdot \varepsilon^t$

The function ε^t is not suitable for $U(t)$ at the beginning of the universe, however, because its value at $t=0$ is not zero. In fact it is zero only at $t = -\infty$. A function that might seem usable, however, would be

(2-9) $U(t) = 0$ $t < 0$ and $t = 0$

       $U(t) = \varepsilon^t - 1$ $t > 0$

$$= t + \frac{t^2}{2!} + \frac{t^3}{3!} + \cdots$$

7

which does have zero value at $t=0$ and otherwise meets the derivatives requirement sufficiently.

Cases (b) – (e): Functions Satisfying (2-5)

Turning to functions that meet the requirement that the second derivative equal the original function per equation *2-5* there are four such functions.

(2-10)
$$\text{Case (b):} \quad U(t) = 1 + \frac{t^2}{2!} + \frac{t^4}{4!} + \cdots$$

(2-11)
$$\text{Case (c):} \quad U(t) = 1 - \frac{t^2}{2!} + \frac{t^4}{4!} + \cdots$$

(2-12)
$$\text{Case (d):} \quad U(t) = t + \frac{t^3}{3!} + \frac{t^5}{5!} + \cdots$$

(2-13)
$$\text{Case (e):} \quad U(t) = t - \frac{t^3}{3!} + \frac{t^5}{5!} + \cdots$$

These five candidate functions can be described and summarized as their exponential equivalents as in Figure 2-2, below.

Case	Function	Name of Function	Candidate U(t)
(a)	ε^t	Natural exponential	$\varepsilon^t - 1$
(b)	$\dfrac{\varepsilon^t + \varepsilon^{-t}}{2}$	Hyperbolic cosine	$\text{Cosh}(t) - 1$
(c)	$\dfrac{\varepsilon^{i \cdot t} + \varepsilon^{-i \cdot t}}{2i}$	Cosine	$\text{Cos}(t) - 1$
(d)	$\dfrac{\varepsilon^t - \varepsilon^{-t}}{2}$	Hyperbolic sine	$\text{Sinh}(t)$
(e)	$\dfrac{\varepsilon^{i \cdot t} - \varepsilon^{-i \cdot t}}{2i}$	Sine	$\text{Sin}(t)$

Figure 2-2

The relationships in the table can be verified by substitution using the formula for ε^t as given in equation *2-6*, above. Cases *(b)* and *(c)* have the same problem that case *(a)* had, that the value of $U(t)$ is not zero at $t=0$. Just as with case *(a)*, they would appear to become satisfactory if a constant, *1*, is subtracted from each of them.

These candidates all satisfactorily meet the requirement for a continuous family of derivatives so that the kind of unacceptable problem as encountered in the example of

$U(t)=t^2$ at the beginning of this discussion is avoided. That is, all derivatives are finite. But, there are other requirements that the successful $U(t)$ function must meet.

b. *Using the Remaining Criteria to Select U(t)*

Two other criteria must be met by the successful candidate function or functions:

- the function must not be open-ended, that is it cannot ever have an infinite amplitude, and

- the function must smoothly match the $U(t)=0$ condition at $t=0$.

The first criterion eliminates cases *(a)*, *(b)* and *(d)* each of which goes to an infinite value of $U(t)$. To satisfy the second criterion the tangent to $U(t)$ at $t=0$ must be identical to the tangent to the function for $t < 0$, which is the horizontal $t\text{-}axis$. The condition is satisfied if the first derivative of $U(t)$ equals *zero* at $t=0$. Only cases *(b)* and *(c)* meet that requirement.

Therefore, the resulting form of $U(t)$, the only acceptable form, the only one that meets all of the requirements, is case *(c)*,

$$(2\text{-}14) \quad U(t) = [Cos(t) - 1] \qquad t > 0 \text{ and } t = 0$$

$$U(t) = 0 \qquad t < 0.$$

which is identical in form to the more usual and convenient equation *2-15*.

$$(2\text{-}15) \quad U(t) = U_0 \cdot [1 - Cos(2\pi \cdot f \cdot t)]$$

in which an amplitude parameter, U_0, and a frequency parameter, *f,* have been added.

That the only possible form for the manner in which the universe began is a sinusoidal oscillatory form would seem to be very appropriate. Oscillations, waves, are ubiquitous in our universe from oceans, violin strings and pendulums to sound, light and electron orbits. That statement can also be validly inverted: Oscillations and waves are ubiquitous in our universe because the universe began from an initial such oscillatory form.

Every oscillation that we know in nature exhibits, and the very theory of oscillations in the abstract requires, that the oscillation consist of two aspects storing and exchanging the energy of the oscillation back and forth by means of a "flow". (With one aspect varying in oscillatory fashion then when that aspect decreases there must be some "place" for its energy to go, a place in which it is stored until it reappears in that aspect when it increases again. It cannot completely disappear or be lost because the oscillation would die. That "place" is the oscillation's second aspect and it obviously must vary in a manner related to the first aspect's variation, but with its energy storage in opposite phase.

A pendulum, for example, oscillates by the motion (flow) of its swinging mass between peak height in the gravitational field (potential energy) at each end of the swing and peak speed of motion (kinetic energy) at the mid-point between the ends of the swing. Then, what is the "flow" of the original oscillation at the start of the universe ? We do not know and likely will never know but we can give it a name, *Medium*, and we can investigate its characteristics and nature.

9

Such was the oscillation at the beginning of the universe except that at the first half cycle the energy was in only one form increasing from zero to its maximum. Then the second form began, similarly from zero to maximum, receiving and storing the energy of the first form as that gradually decreased in the second half cycle.

2 - THE PROBLEM OF CONSERVATION – "SOMETHING FROM NOTHING"

At this point, that is the universe having started from absolute nothing as an oscillation having the form of equation *2-15*, the maintaining of conservation, the avoiding of getting something from nothing, clearly could only happen in one manner:

> There simultaneously had to have arisen an identical-in-form but opposite-in-amplitude oscillation so that the pair balanced out to the original net nothing, as in equation *2-16*.

$$(2\text{-}16) \quad U(t) = \pm\, U_0 \cdot [1 - \mathrm{Cos}(2\pi \cdot f \cdot t)]$$

There is no other way that violating the assured principle of conservation could have been avoided. The universe exists. It had to come into being from a prior nothing. That had to happen while avoiding an infinity of rate of change. Conservation had to be maintained. The universe began with the oscillation of equation *2-16*.

3. THE PROBLEM: WHY THAT OSCILLATION BEGAN AND WHAT IT WAS

a. Why That Beginning happened

A duration is the period of time that a particular state or set of conditions persists. The duration is terminated by a change, which change also initiates a new duration. In the universe change is ubiquitous. It is the constant and continuous stream of change that makes durations mensurable. Before the beginning of the universe a duration was in process even though it was not mensurable. The beginning of the universe was the first change ever and it terminated the original primal duration of absolute nothing.

The probability of the happening of such an event is extremely small. But the event was / is not impossible. Furthermore, in the absence of that event occurring there was an extremely large duration of opportunity in which that extremely small probability could operate. In the absence of the beginning the original duration would have been infinite and that infinite opportunity operated on by minute, but non-zero, probability results in absolute certainty. The beginning of the universe could not avoid eventually happening.

b. What That Beginning Oscillation Was

The starting point is the assumption that, when the primal nothing changed as a probabilistically inevitable interruption of what would otherwise have been an infinite duration of the primal nothing, the simplest or minimum conservation-maintaining interruption that could occur is what occurred. There are two reasons for this. Occam's Razor, calls for the simplest hypothesis as the most likely. More importantly, or perhaps the same thing, if an essentially spontaneous and extremely low probability event is to occur solely as an interruption of the duration of an otherwise absolute nothing, then very little interrupting event is needed; the barest minimum of something is sufficient to interrupt, to be a change in absolute nothing. There is no call, no reason for anything more. So, while the interruption could have been otherwise, it was probably as simple and minimum as possible.

Size or amount of time are of no meaning here because there is nothing to which they can be compared or by which they can be measured. Whatever amount of change occurred is what occurred. Whatever time it took, or went on for, whatever its oscillatory frequency was, is what happened. Twice as much or half as much have no meaning.

The following conclusions about the initial oscillatory $\pm U_0 \cdot [1 - Cos(2\pi \cdot f \cdot t)]$ form can now be reasonably obtained:

- clearly the universe of today must be an on-going evolved consequence of its beginning, of the initial oscillatory form;

- the frequency, f, of the sinusoidal oscillation was, and is, very large; and

- the nature of the change is one of concentration or density of the something that is oscillating.

- the oscillation was spherical, radially outward in all directions from its origin, because there was nothing to constrain it otherwise.

The frequency would have to be either very large or very small -- high enough so that it is not detected or noticed by us in every day life or so low that it appears to us as no change at all in our experience.

It has already been noted that the fact that the only possible form for the manner in which the universe began is a sinusoidal oscillatory form is very appropriate because oscillations, waves, are ubiquitous in our universe from oceans, violin strings and pendulums to sound, light and electron orbits. And it has been noted that that statement can be validly inverted: oscillations and waves are ubiquitous in our universe because the universe began from an initial such oscillatory form.

If the frequency of the initial oscillation were so small that it appears to us as no change at all it would completely eliminate oscillations playing any significant part in the behavior of the universe as we know it. Therefore, the frequency must have been very large, so rapid compared to our perception that we do not notice the oscillation at all.

The change can hardly be one of gross size if it is going on right now at high frequency as has just been concluded. One can conceive of the fundamental "substance", the "something" of the universe flashing into and out of existence from a zero to a maximum density or concentration in an oscillatory fashion at a rate so high that we neither detect nor notice it at all. But, it is not possible to entertain a concept of reality flashing from zero to full size, a size that includes ourselves and our environment, in such a fashion.

Actually, the reality that we know is not "flashing into and out of existence ...." Our reality is more the oscillation itself than what is oscillating and the continuing oscillation is our steady, constant reality.

All of the discussion so far must apply to the "negative" oscillation, $-U(t)$, exactly as to the "positive" oscillation $+U(t)$ because the exact same reasoning as for $+U(t)$ applies to $-U(t)$ and, after all, they are not distinguishable in the discussion. The terms "+" and "−" are merely terms of convenience for two equal form opposite

11

magnitude unknown things. We probably tend to think of our universe as the "+", but that is meaningless and irrelevant. There can be no objective designation of $+U(t)$ and $-U(t)$, no way to identify one versus the other. Both had to appear and our universe cannot avoid being the evolved result of both.

The universe that we know and exist in is the combined integrated result of both $+U(t)$ and $-U(t)$. The "+" and "-" electric charges of our universe [in both matter as for example in protons and electrons and in anti-matter as for example in negaprotons and positrons] must derive from that aspect of the beginning. (It is interesting to observe, also, that our universe being the integrated result of an initial beginning and its opposite relates to (presumably is the underlying cause of) the dialectical nature of reality, the ying and yang of oriental philosophy.)

The question of what the *Medium* is can only be answered in terms of its characteristics, what it does and how. Its characteristics are:

- a continuous entity, not a mass of "particles" nor anything having parts,

- simple and uniform throughout,

- of minimum tangibility or substantiality, not unlike the actuality of what we designate as "field" [electric, gravitational, etc].

4. THE PROBLEM: WHY DID THE EFFECTS OF EQUATION 2-16 NOT PROMPTLY CANCEL AND ON-GOING ABSOLUTE NOTHING RESUME ?

This is resolved in detail in Appendix A, *Why No Immediate Mutual Annihilation*. Briefly, the initial structure was so unstable that it promptly exploded in that which we refer to as the "Big Bang" before annhilation could occur.

5. THE PROBLEM: IT HAS BEEN THOUGHT THAT THE UNIVERSE HAD TO START AT A POINT. HOW COULD A POINT DELIVER A WHOLE UNIVERSE?

The sole reason for positing a point origin was to avoid an initial infinite rate of change. The gradualness of the $[1 - Cosine]$ form resolves the problem of avoiding an infinite rate of change so that a point origin is no longer required.

The Big Bang "event horizon" problem and its relation to the development of variety in the universe has led to the hypothesis that there was an initial brief period of extremely rapid expansion called "inflation". That hypothesis has no supporting cause nor mechanism except its role in meeting the "event horizon" problem.

But with the need for a point origin eliminated the origin can have started per equation $2-16$ at any size. There was no un-accounted-for period of "inflation". From estimates calculated of the number of particles in today's universe it has been determined that the initial, at the very first instant, the already "inflated"–size universe began. It was a highly concentrated volume of all of the mass and energy of the universe of about $40,000\ km$ radius.

That size is in terms of today's sizes. For that event at that time specific size is meaningless because there was nothing else to compare it to.

Thus the interruption of the original primal absolute nothing that gave us our universe was the starting of an *oscillation* that was *spherical*, present to us at a very high frequency and of $\pm U_0 \cdot [1 - Cos(2\pi f \cdot t)]$ form, of the *Density*, as the variation will be hereafter referred to, of the *Medium*, as what it is that is oscillating will be hereafter referred to.

(2-16) $\quad U(t) = \pm U_0 \cdot [1 - Cos(2\pi \cdot f \cdot t)]$

From the Origin to the Complex Universe

Section 2, *The Origin of Space and Matter* resolved the origin of the universe as follows.

> The universe exists. It could only come from a prior nothing. That had to happen while avoiding an infinite rate of change. Conservation had to be maintained. Therefore the original oscillation, equation (2-16): $U(t) = \pm U_0 \cdot [1 - Cos(2\pi \cdot f \cdot t)]$
> Then, how did that develop
> into the complex matter we experience today?

The hypothesis is that the interruption that started our universe, the interruption of what would otherwise have been an infinite duration of the primordial absolute nothing, an interruption because an essentially infinite amount of opportunity operated on a non-zero though minute probability, was the starting of a matched pair of spherical oscillations:

- Present to us at a very high frequency,
- Of the general *[1 - Cosine]* form, and
- Together equal to the original nothing because of having
  matching amplitudes $+U_0$ and $-U_0$.

That analysis yielded an initial event, the origin oscillations, as in Figure 3-1. [All of the unavoidably planar depictions of the spherical oscillations are of the spherical phenomenon, interpretable as a radial versus time depiction.]

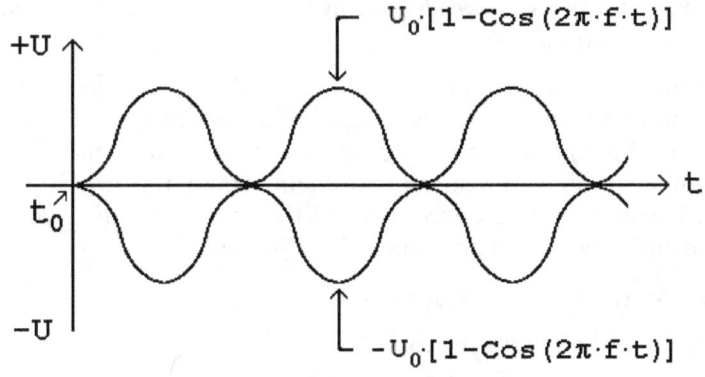

Figure 3-1

15

HOW THE ORIGINAL OSCILLATIONS BECAME THE UNIVERSE

Examination of the waveform of Figure 3-1 reveals two problems. One, that it is an immediate mutual annihilation, will be dealt with shortly below. Of concern now is that an infinite rate of change still remains; the envelope of the oscillation has an infinite rate of change at $t=t_0$ as can be seen in Figure 3-2, below, which displays the envelope.

Viewed in a mathematical or graphical sense without any consideration of the physical reality represented, the envelope discontinuity at $t=t_0$ is not a difficulty. The only quantity that actually exists and is varying is the overall $U(t)$. The envelope is merely our perception of a characteristic of the waveform. The actual varying quantity, per Figure 3-1, has no discontinuity at $t=t_0$

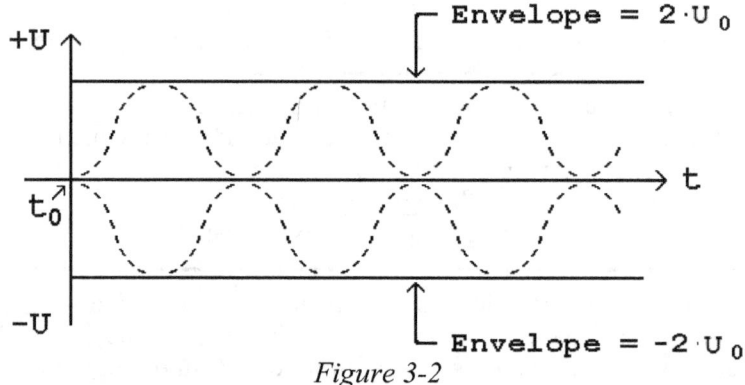

Figure 3-2

However, looking at the situation in a physical sense rather than purely mathematically, such oscillations as depicted in Figure 3-1 are all that there is to account for the effects which we call *energy, mass* and *charge*. Therefore, this *energy / mass / charge / oscillation* is something other than nothing. It is a physical reality that did not exist prior to the Origin. It can no more leap from zero to a finite non-zero amount than could the original $U(t)$ so leap.

That infinite rate of change in the amount of *energy / mass /charge* at $t=t_0$ is no more acceptable than was the infinite rate of change encountered in the original analysis of the beginning and it must be corrected by the same kind of reasoning as was then pursued: the envelope, also, had to originate as a *[1 - Cosine]* form of oscillation, which is the only form that avoids an infinite rate of change and matches the requirements of the situation.

That original envelope oscillation was at a lesser frequency than the original wave by the definition of a waveform envelope. If it were at a greater frequency then the roles (envelope and wave) would be reversed. If it were at the same frequency it would not act as an envelope and the infinity problem would remain. If we designate the envelope frequency as f_{env} and the frequency of the wave oscillation within the envelope as f_{wve} then the envelope would be of the following form.

(3-1) $\qquad U_{env} = [1 - Cos(2\pi \cdot f_{env} \cdot t)]$

The wave is, as before, of the form

(3-2) $\qquad U_{wve} = \pm U_0 \cdot [1 - Cos(2\pi \cdot f_{wve} \cdot t)]$

and the envelope modulating the wave is then

16

(3-3) $U(t) = [U_{env}] \cdot [U_{wve}]$

$\qquad = \pm U_0 \cdot [1 - \cos(2\pi \cdot f_{env} \cdot t)] \cdot [1 - \cos(2\pi \cdot f_{wve} \cdot t)].$

That waveform appears in Figure 3-3.

However, the form of *U(t)* of equation *3-3* and Figure 3-3 still does not resolve the problem of an infinite rate of change at t_0. The *[1 - Cosine]* envelope is itself an oscillation that begins at t_0 with a sudden step from zero to its full amplitude. Figure 3-3 shows the first *2* cycles of the envelope oscillation, which if only the envelope is considered, is a simple oscillation at the envelope frequency, even though visually, in the Figure, it is only the trace of the peaks of the overall complex oscillation.

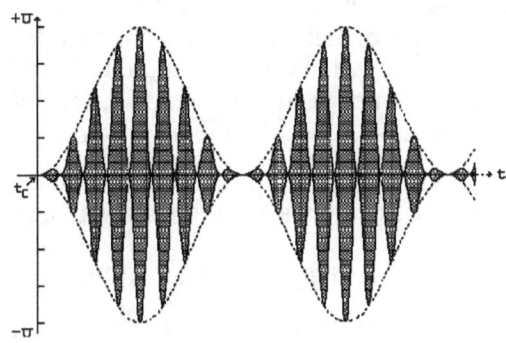

Figure 3-3

It is *energy / mass / charge* that begins suddenly in its full amount at t_0 just as, in Figure 3-1, the oscillation of equation *3-1* begins at t_0. Therefore, it is again necessary to introduce an envelope of *[1 - Cosine]* form to prevent the infinite rate of change at t_0 in the prior envelope. That correction will in turn require still another such correction and so *ad infinitum*. An (apparently at this point) infinite string of envelopes thus results as a necessity of the situation.

The resulting *U(t)* then is

(3-4)

$$U(t) = \pm U_0 \cdot \prod_{i=1}^{i=\infty} \left[[1 - \cos(2\pi \cdot f_{env_i} \cdot t)] \right] \cdot \quad \cdots$$

$$\cdots \quad \cdot \left[[1 - \cos(2\pi \cdot f_{wve} \cdot t)] \right]$$

where the $\prod$ symbol (a large π, Greek "p") means the product of the indicated factors.

While an envelope frequency must be less than the frequency of the wave that it modulates so that the various f_{env} must be less than f_{wve}, each successive envelope may be at the same frequency, as the prior. The reason is as follows.

If each envelope frequency must be different then each must be at least slightly smaller than the prior. With an infinite set of envelopes and only the frequency range from slightly less than that of the wave down to slightly above zero being available each successive envelope could only be at an infinitesimally lower frequency than its predecessor in any case. Infinitesimally less is essentially the same as identical.

Then how did other than an infinite string of envelopes come about ?

Each additional envelope factor in equation *3-4* results in a higher frequency content in the overall expression. That is, as each envelope is added the expansion of the exponentiated cosines expression into a sum of individual frequency cosine terms becomes longer and acquires higher frequency terms. But, the oscillation could not have had an actual component at infinite frequency. The real universe original *U(t)* had an enormous set of envelopes but not an infinite set; they were "cut off" at some point.

The *Medium* of these oscillations being the only reality and, therefore, being what sets the limit on the speed of light with which we are familiar, the *Medium* also sets a limit on the highest frequency / lowest wavelength waves that can propagate. As a result the series of envelopes, of factors in equation *3-4*, was limited to some finite but quite large amount. (See Appendix B, *The Limitation of the Original Envelopes*).

This yields a revised *U(t)*, the original oscillation, the Cosmic Egg, as equation *3-5*, below. N_0 is the number of envelopes, all at the same frequency, f_{env}.

$$(3-5) \qquad U(t) = \pm U_0 \cdot \left[1 - \cos\left[2 \cdot \pi \cdot f_{env} \cdot t\right]\right]^{N_0} \cdot \left[1 - \cos\left[2 \cdot \pi \cdot f_{wve} \cdot t\right]\right]$$

The waveform $[1 - \cos(x)]^n$ converges to an increasingly narrower peak as n increases, Figure 3-4, below. For very large n, that is very large N_0 of equation *3-5*, the converging of the waveform into a single narrow peak proceeds to a momentary "spike" per cycle. Figure 3-5, below, shows the appearance of the waveform for extremely large n, that is for $n = N_0$ - what the waveform of the original "Cosmic Egg", the start of our universe, "looked like". (N_0 is found further below to be about 10^{84}.)

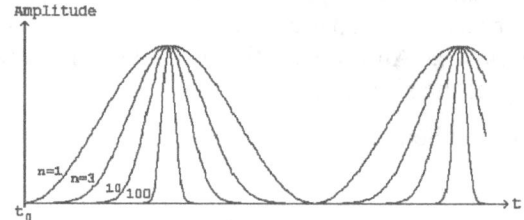

Figure 3-4 $[1 - \cos(x)]^n$ *For* $n = 1, 3, 10, 100$

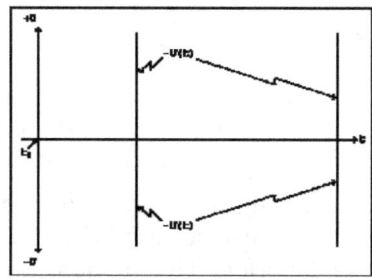

Figure 3-5
The U(t) "Cosmic Egg" WaveForm

18

This discussion of $U(t)$, the original oscillation the start of which was the start of the universe, has dealt so far only with the problems of the Origin, the problems of the transition from nothing to something. The something was, of course, the first instant of the entire universe. As such it must have contained in itself all of the *mass / energy / positive and negative charge* of the universe.

Figures 3-1, 3-3, and 3-5 all indicate that the original pair of oscillations, $+U$ and $-U$, should have immediately mutually annihilated, canceled out, reverted to the primal nothing. But, clearly that did not happen. The only explanation of that not happening is that each was unstable, so unstable that they exploded more immediately than they were able to mutually annihilate. They immediately proceeded to an immense explosion of energy and pieces of their oscillation, the event now called the "Big Bang". See Appendix A, *Why No Immediate Mutual Annihilation.*

In terms of the $U(t)$ as depicted in Figure 3-5, the so immediate explosive decay undoubtedly occurred after only a minute portion, an infinitesimal portion, of the very first cycle had passed. It had to have been long before the first "spike". In that sense the initial event was very small, tenuous, hardly more than nothing because the instantaneous amplitude of $U(t)$ at that moment (the height of the curve above zero at that moment long before the first "spike") was also infinitesimal. It was hardly more than, essentially zero.

In that sense, the way that the universe started at all becomes a little more comprehensible. To avoid an infinite rate of change there was essentially almost no difference between "nothing", on-going absolute nothing, and the first infinitesimal moment of the original $U(t)$, the original oscillation.

Yet, it contained the entire universe.

THE FORM OF MATTER AS GENERATED BY THE "BIG BANG"

What did the "Cosmic Egg" explode into ? It could only explode into pieces of what it was made of, pieces of *[1 – cosine]* form spherical oscillations, pieces like equation *2-16*, above.

Each oscillation is three-dimensional, thus spherical, because three dimensions is the minimum number that can involve space part of which is not its own boundary.

But, what did the "Cosmic Egg" explode into ? It primarily exploded into what we know our universe to mainly consist of: myriad protons - Hydrogen atom nuclei, and myriad electrons - maintaining overall charge neutrality with the protons, and the antimatter forms of both, negaprotons and positrons – maintaining conservation.

[Those might also be expected to have mutually annihilated but did not. Their survival rather than annihilation is analyzed in full in Appendix A, *Why No Immediate Mutual Annihilation.* Suffice it here to observe that each product piece was initially ejected radially outward at extreme velocity and energy, on paths slightly diverging, such that initially annihilations could not occur.]

Then, what was the nature, the form of those product pieces that the "Cosmic Egg" exploded into ? Because of the two frequencies of $U(t)$, f_{wve} and f_{env}, and that the explosion source was of two equal but opposite polarities, $+U_0$ and $-U_0$, the "Big Bang" resulted in myriad pieces of four different forms of *[1 – cosine]* form spherical oscillations , equations *3-6* .

(3-6) $U_{Form\ 1}(t) = +U_c \cdot [1 - Cos(2\pi \cdot f_{wve} \cdot t)]$ the proton

        $U_{Form\ 2}(t) = -U_c \cdot [1 - Cos(2\pi \cdot f_{env} \cdot t)]$ the electron

        $U_{Form\ 3}(t) = -U_c \cdot [1 - Cos(2\pi \cdot f_{wve} \cdot t)]$ the anti-proton

        $U_{Form\ 4}(t) = +U_c \cdot [1 - Cos(2\pi \cdot f_{env} \cdot t)]$ the anti-electron

Each of those has a specific value of its mass. Per the data provided by NIST, the National Institute of Standards and Technology those masses are:

(3-6a)   ▪ the proton and the antiproton $m_p = 1.672\ 621\ 898 \cdot 10^{-27}$ *kg*

        ▪ the electron and the anti-electron $m_e = 9.109\ 383\ 56 \cdot 10^{-31}$ kg.

Using the mass-energy relationship, $m \cdot c^2 = h \cdot f$ the frequency, f, of those particles can be calculated. Those frequencies are:

(3-6b)   ▪ the proton and anti-proton: $f_{wve} = 2.268,731,818 \cdot 10^{23}$ hz

        ▪ the electron and anti-electron: $f_{env} = 1.235,589,965 \cdot 10^{20}$ hz.

Finally, the mass of those four fundamental particles having now been resolved, their electric charge remains. They all have the same magnitude of their oscillation, $|U_c|$, which by default is the magnitude of their electric charge. [U_c is the particle oscillation amplitude per equation 3-6. U_0 is the original pre-explosion oscillation amplitude.] The magnitude of the oscillation is in two opposite polarities; therefore clearly, where q is the fundamental electric charge per NIST, then:

(3-7) $q = 1.602,176,621 \times 10^{-19}$ c

   $+U_c = +q$

   $-U_c = -q$

Judging by its result, the "Cosmic Egg" was not unlike an immense atom, a very unstable immense atom [as are all of the atomic species of atomic number exceeding 83 which the cosmic egg would have immensely exceeded]. Its "Big Bang" was a kind of explosive nuclear radioactive decay ultimately ending in the myriad stable elements of today's Periodic Table plus those with half lives long enough to be in detectable quantities today. Such decays follow a chain:

   - From a heavy and complex composition,

   - To various multiple less heavy less complex product pieces,

       · · · · ·

   - Until they arrive at many multiple stable forms.

The vast majority of those resulting stable forms are the protons and electrons of the material world and their anti-particles. They are of the equation *3-6* form spherical oscillation, and will be referred to as *Spherical-Centers-of-Oscillation* or as *particles*

The rates of the decays are exponential, the decay [varying from some extremely rapid to some extremely slow] is described in terms of a "half life", the time it takes for half of the original material's decays to take place. Some of those "multiple less heavy less complex product pieces" having long half lives are present to us still today still decaying as what we term "radioactive" species.

20

The actions of the various stable atomic forms are primarily: electrostatic per Coulomb's Law, electromagnetic per Ampere's Law, and gravitational per Newton's Law. Those are treated in detail as well as the process of radioactive decay in the Book *On the Nature of Matter* published by The-Origin Foundation, Inc., at its website www.The-Origin.org.

> Of interest here is the nature of gravitation which develops from a particular behavior of the *Spherical-Centers-of-Oscillation,* which behavior is developed and analyzed in the following Section 4, The *Outward Flow from all Particles*.

The Outward Flow from All Particles

> The gravitational and electrical interactions of particles requires <u>communication</u> between them. That comes about as follows.

The gravitational effect between two gravitating objects is the net combined vector effect of myriad individual gravitating particles of one of the two objects interacting gravitationally in Particle-on-Particle pairs with myriad individual gravitating particles of the other of the two objects.

Particle's Propagated Outward Flow

Each and every gravitationally attrac**ting** *Spherical-Center-of-Oscillation* must communicate to each gravitationally attract**ed** particle its "message": the direction from the attract**ed** particle to the attrac**ting** one and the magnitude of the attrac**ting** particle's gravitational attraction. That task is assigned by contemporary physics' theory to a *gravitational field*, a vector field that is an assignment of a direction of action and its magnitude to each point in a region of space.

However, that designation of the field, while facilitating the description of the action fails to explain the cause, the mechanism of the field and thus fails to explain or account for the action at issue. It also fails to account for the time delay due to the limitation of the speed of light that must exist between a change at the attrac**ting** particle and its effect at the attract**ed** particle.

Something flowing is required, something flowing at the speed of light, continuously, carrying the direction and magnitude information, spherically outward, from every gravitating *Spherical-Center-of-Oscillation* to every other *Spherical-Center-of-Oscillation*. That flow is contemporary physics theory's gravitational field.

For such a flow to persist there must be a supply of that outward flowing substance in every particle. And, for that flow to have persisted the billions of years since the "Big Bang" that "supply" must be an extremely concentrated reservoir of that which flows outward [concentrated relative to the outward flow].

23

THE PARTICLE "CORE"

Consider a small individual particle such as a proton. Newton's law of gravitation expressed in terms of the masses, m_{source} and $m_{acted-on}$, and with both sides of the equation divided by $m_{acted-on}$ is, of course,

$(4-7)$ $\qquad a_{grav} = G \cdot \left[\dfrac{m_{source}}{d^2} \right]$

However, mass and energy are equivalent, so that [using c = light speed and h = Planck's constant] a mass, m, is proportional to a frequency, f, that is characteristic of that mass. That is

$(4-8)$ $\qquad m \cdot c^2 = h \cdot f \quad$ or $\quad f = [c^2/_h] \cdot m$

so that the m_{source} of equation $4-7$ has a corresponding equivalent frequency, f_{source}.

That being the case, the gravitational acceleration, a_{grav}, can be expressed in terms of that frequency as the change, Δv, in the velocity, v, of the attracted mass per time period, T_{source}, of the oscillation at the corresponding frequency, f_{source}, as follows.

$(4-9)$ $\qquad a_{grav} = \Delta v / T_{source} = \Delta v \cdot f_{source}$

It can then be reasoned using equation $4-9 =$ equation $4-7$ as follows .

$(4-10)$ $\qquad a_{grav} = \Delta v \cdot f_{source} = G \cdot \left[\dfrac{m_{source}}{d^2} \right]$

Equation $4-11$, below, is obtained by using that frequency is proportional to mass. With f_p and m_p as the proton frequency and mass then $f_{source} = [m_{source} / m_p] \cdot f_p$.

$(4-11)$ $\qquad \Delta v \cdot \left[\dfrac{m_{source}}{m_p} \right] \cdot f_p = G \cdot \left[\dfrac{m_{source}}{d^2} \right]$

Rearranging and canceling m_{source} on both sides of the equation,

$(4-12)$ $\qquad \Delta v = \dfrac{G \cdot m_p}{d^2 \cdot f_p} \quad$ per cycle of f_{source}.

Then substituting, per equation $4-8$, $m_p = [h \cdot f_p] / c^2$,

$(4-13)$ $\qquad \Delta v = \left[\dfrac{G}{d^2 \cdot f_p} \right] \cdot \left[\dfrac{h \cdot f_p}{c^2} \right]$

$\qquad\qquad = \dfrac{G \cdot h}{d^2 \cdot c^2} \quad$ per cycle of f_{source}.

The Planck Length, l_P, is defined as

$(4-14)$ $\qquad l_P \equiv \left[\dfrac{h \cdot G}{2\pi \cdot c^3} \right]^{\frac{1}{2}} \quad$ so that $\quad G = \left[\dfrac{2\pi \cdot c^3 \cdot l_P^2}{h} \right]$

24

Substituting G as a function of the Planck Length from equation *4-14* into G as it is in equation *4-13*, the following is obtained.

(4-15) $$\Delta v = \left[\frac{2\pi \cdot c^3 \cdot l_P^2}{h}\right] \cdot \left[\frac{h}{d^2 \cdot c^2}\right]$$

$$= c \cdot \frac{2\pi \cdot l_P^2}{d^2} \quad \text{per cycle of } f_{source}.$$

This result states that:

- the velocity change due to gravitation, Δv,
- per cycle of the attracting mass's equivalent frequency, f_{source}, which quantity, $\Delta v \cdot f_{source}$, is the gravitational acceleration, a_{grav},
- is a specific fraction of the speed of light, c, namely the ratio of:
  - 2π times the Planck Length squared, $2\pi \cdot l_P^2$, to
  - the squared separation distance of the masses, d^2.

That squared ratio is, of course, the usual inverse square behavior.

This also means that at distance $d = \sqrt{2\pi} \cdot l_P$ from the center of the source, attracting mass, the acceleration, Δv, per cycle of that attracting mass's equivalent frequency, f_{source}, is equal to the full speed of light, c, the most that it is possible to be. In other words, at that [quite close] distance from the source mass the maximum possible gravitational acceleration occurs. That is the significance, the physical meaning, of l_P or, rather, of $\sqrt{2\pi} \cdot l_P$.

The physical significance of $\sqrt{2\pi} \cdot l_P$ is that it sets a limit on the minimum separation distance in gravitational interactions and it implies that a "core" of that radius is at the center of fundamental particles having rest mass. That is, equation *4-15* clearly implies that it is not possible for a particle having rest mass to be approached closer than that distance.

That physical significance of $\sqrt{2\pi} \cdot l_P$, is so fundamental to gravitation and apparently to particle structure, that it more truly represents a fundamental constant than does l_P. For those reasons that length should replace l_P as a fundamental constant of nature as follows.

(4-16)   The fundamental distance constant, δ

$$\delta^2 \equiv 2\pi \cdot l_P^2$$

$$\delta = 4.051,34 \times 10^{-35} \text{ meters}$$

Equation *4-15* then becomes equation *4-17*.

(4-17) $$\Delta v = c \cdot \frac{\delta^2}{d^2} \quad \text{per cycle of } f_{source}$$

a quite pure and precise statement of gravitation: that gravitation is a function of the speed of light, c, and the inverse square law, in the context of the oscillation frequency, f_{source}, corresponding to the attracting, source body's mass.

25

That makes clear that an oscillation is an integral part of gravitation as should be the case because gravitation is an action between particles having mass, which are the *Spherical-Centers-of-Oscillation* products of the "Big Bang".

Having now just determined:

- That δ sets a limit on the minimum separation distance in gravitational interactions and therefore that a "core" of that radius is at the center of fundamental particles, and
- That an extremely concentrated reservoir supply of that which is flowing outward is required at the center of all particles to support the billions of years of their outward flow;

Therefore:

- The reservoir is the spherical "core" of radius δ at the center of all particles;
- That it is impenetrable is because of its immense density concentration [billions of years worth of flow of the flow substance [*Medium*] in the minute ($\delta = 4.05134 \times 10^{-35}$ *meters* radius spherical core) of every particle having rest mass], and.
- The *Spherical-Center-of-Oscillation* is a spherical oscillation of that immensely concentrated flow substance, *Medium*.

Then, what "contains" that core's supply or why doesn't it all just quickly "slosh" out and be gone ? It is trying to do just that, to "slosh" out, as hard as it can. It cannot help propagating outward because it has no container. But it can only propagate outward at the limiting rate determined by its surface area, $4 \cdot \pi \cdot \delta^2$ and the fastest speed possible for flow, the speed of light, c. Thus is the *Propagated Outward Flow*.

The Speed of the Flow – The Speed of Light

Every oscillation that we know in nature exhibits, and the very theory of oscillations in the abstract requires, that the oscillation consist of two aspects of the oscillating substance storing and exchanging back and forth the energy of the oscillation [e.g. pendulum position and velocity or electric potential and current]. With one aspect varying in oscillatory fashion then when that aspect decreases there must be some "place" for its energy to go, a place in which it is stored until it reappears in that aspect when that aspect increases again. It cannot completely disappear or be lost because the oscillation would die. That "place" is the oscillation's second aspect and it obviously must vary in a manner related to the first aspect's oscillatory variation with its energy storage in opposite phase.

As is the case for electric inductance and capacitance determining the speed of propagation along a transmission line, μ_0 and ε_0 determine the speed of the *[1 - Cosine]* form oscillation propagation by setting the two aspects of the oscillation in which they are involved, the aspects between which the oscillation energy exchanges back and forth.

But, when the original oscillation came into existence it did so in absolute nothing. There was no "free space" with μ_0 and ε_0. There was nothing but the original oscillation and nothing at all beside that. And, after the immediate explosion into all of the particles of the universe, each of those particles was sending its *Propagated Outward Flow* <u>into nothing, into emptiness.</u>

26

Where did the *Propagated Outward Flow*'s μ_0 and ε_0 come from? The only thing they could have come from was the original oscillation. There is no other possible source because everything else was absolute nothing. The μ_0 and ε_0 are inherent in the substance of the oscillation, which means, μ_0 and ε_0 are also inherent in the outward propagation. Each particle's *Propagated Outward Flow* of *Medium* contains and carries its own μ_0 and ε_0.

Having established the supply of *Medium* [flow substance] and its on-going *Propagated Outward Flow* serving the role of gravitational field as a property of every particle exhibiting rest mass, the question arises, "What of the electric field, much stronger than gravitation and co-present with gravitational field whenever the gravitating particle has electric charge ?"

Just as is the case for gravitation as presented above, every particle having electric charge must communicate its similar "message" to every other such particle. That requires something flowing outward at the speed of light continuously, carrying the direction and magnitude information, spherically outward, from every electrostatic *Spherical-Center-of-Oscillation* to every other *Spherical-Center-of-Oscillation*. That flow-communication is the electric field, an active process not a static state.

The theory of an *electric field*, just as with that of a *gravitational field*, above, while facilitating the description of the action fails to explain the cause, the mechanism of the field and thus fails to explain or account for the action at issue. It also fails to account for the time delay due to the limitation of the speed of light that must exist between a change at the attract**ing** particle and its effect at the attract**ed** particle

Two such simultaneous flows, gravitational and electric, and two supporting reservoirs supplying the flows, is clearly untenable. There can only be one reservoir in each particle's "core" and one resulting *Propagated Outward Flow* producing both the gravitational action and the electric action if for no other reason than because two supply reservoirs would mutually interfere with a spherically outward flow of each.

The one sole flow performs both the Coulomb and the gravitational action: the Coulomb mediated by the flow amplitude and the gravitational mediated by the flow's frequency or repetition rate. The dual functions of the sole single outward flow from all particles is important in the development of control of gravitation later in this work. It means that electro-magnetic light and gravitation are closely related through their shared common *Propagated Outward Flow*.

See On the *Nature of Matter*, Roger Ellman, The-Origin Foundation, Inc.

SUMMARY FOR SECTION 4 – THE OUTWARD FLOW FROM ALL PARTICLES

The form of matter is not that of the "particles" of classical modern physic's Standard Model. Rather the form of matter is:

- *Spherical-Centers-of-Oscillation*, spherical oscillations of [1 - Cosine] form;

- Propagating spherically outward a continuous oscillatory *Propagated Outward Flow* of *Medium* in [1 - Cosine] form, according to its source *Spherical-Center-of-Oscillation* magnitude, sign, and frequency;

- The speed of the *Propagated Outward Flow*, c, set by the net μ and ε in the *Medium* being propagated;

(4-18)
$$c = \frac{1}{\sqrt{\mu \cdot \varepsilon}}$$

The *Spherical-Center-of-Oscillation* consists of a central "core", a spherical volume of radius $\delta = 4.051,34 \times 10^{-35}$ meters that consists entirely of a high density concentration of the oscillating *Medium*, which propagates outward at an extremely low rate determined by the surface area of the "core" and the radial outward speed of flow of the propagated *Medium*, the speed of light, c.

The Mechanism of Gravitation

How the Vacuum Magnetic Permeability and the Vacuum Electric Permittivity, μ_0 and ε_0, of particles' outward *Flow* produce gravitational attraction.

THE AFFECT OF THE OUTWARD FLOW ON ITS PROPAGATING SOURCE

The oscillating substance in the core of each of the myriad particles is its mass. There is no other place or thing to be the mass of those particles. Therefore the propagating outward *Flow* has momentum, the effect of the product of mass, inherent in the substance of the *Flow*, and that substance *Flow*'s velocity.

The *[1 - cosine]* oscillatory form of the propagated wave is, in effect, a stream of pulses. In the absence of other effects the outward *Flow* is naturally radially outward. While the radially outward *Flow* of the particle's core source of the *Flow* effectively transmits pulses of momentum outward in its *[1 - cosine]* oscillation, that core source of that *Flow* simultaneously experiences radially inward equal but opposite pulses of reaction momentum in accordance with Newton's third law of motion, to every action there is an equal but opposite reaction. In effect the core source is under radial reaction compression. Because that effect is usually radially uniform it produces no net affect on the particle. Figure 5-1.

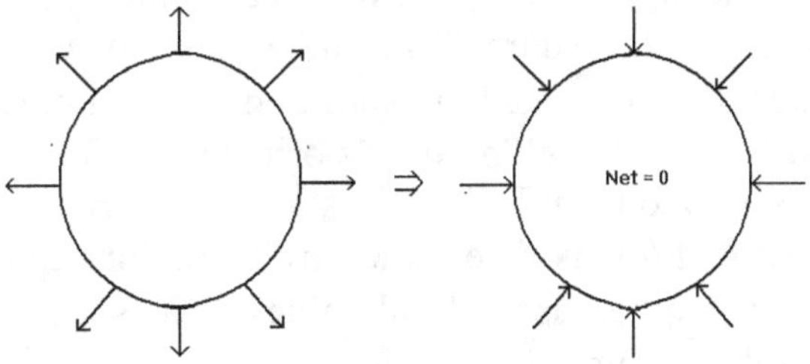

Outward *Flow* Pulses   Cause   Equal but Opposite
Reaction Pulses Inward to no Net Effect on the Particle's Motion

Figure 5-1

31

A PARTICLE'S FLOW ENCOUNTERING ANOTHER PARTICLE

In a universe of the myriad particles resulting from the Big Bang, each of those particles propagating its own outward *Flow* radially in all directions, there are many instances of the *Flow* from one particle [the "source" particle] encountering, running into, the outward *Flow* of another particle [the "encountered" particle]. Because the *Flow*s are spherically outward *Flow*s from spherically oscillating sources, such "source" particle *Flow*s are oscillatory and inverse square reduced in magnitude the farther that their wave front has traveled from its source.

The *Flow* behavior is analogous to that of an electric transmission line where the rate of travel of an oscillation down the line is determined by the time it takes to build up the electric current for each oscillation cycle through each infinitesimal increment of the line's distributed series inductance [L_p] and to build up the electric potential for each oscillation cycle on each infinitesimal increment [C_p] of the line's distributed shunt capacitance. The transmission line speed of *Flow* is determined by the well-established relationship equation *(5-1)*.

(5-1) $$\text{Speed}_{\text{Transmission Line}} = \frac{1}{\sqrt{L_p \cdot C_p}}$$

As presented in the preceding Section 5, the μ_0 and ε_0 are inherent in the substance of the oscillation, which means, μ_0 and ε_0 are also inherent in the outward propagation. Each particle's *Propagated Outward Flow* contains and carries its own μ_0 and ε_0.

For particles' propagating oscillating *Flow* the factor determining its speed of propagation is the time required to build up the *Flow* amount for each oscillation cycle through each infinitesimal increment of the *Flow*'s μ_0 and the *Flow*'s potential for each oscillation cycle on each infinitesimal increment of the *Flow*'s ε_0. But, in radially outward propagating particle's *Flow*, the *Flow* amount is inverse square spread out and the potential likewise both in exactly the same proportion as its μ_0 and ε_0. The ratio of the *Flow* amount to its μ_0 and of its *Flow* potential to its ε_0 remains constant, and so likewise the speed, radially outward, of its propagation, *c*.

Upon encountering another particle that arriving *Flow*'s then μ_0 and ε_0 (scalar quantities not vector and much inverse square reduced) combine with the (full magnitude as in the *Flow* as originated at the "source") μ_0 and ε_0 in the new outgoing propagation of the "encountered" center, the combined μ_0 sum and the combined ε_0 sum each being larger values than in the "encountered" particle's originating *Flow*. The result is that that "encountered" particle's new outward *Flow* is slowed relative to its natural otherwise speed. That is, its speed of *Flow* is determined by a combination of the parameters μ_0 and ε_0 larger than its *Flow*'s otherwise natural values. The speed of *Flow* is determined by the well-established relationship:

(5-2) $$\text{Speed} = \frac{1}{\sqrt{\mu_0 \cdot \varepsilon_0}}$$

32

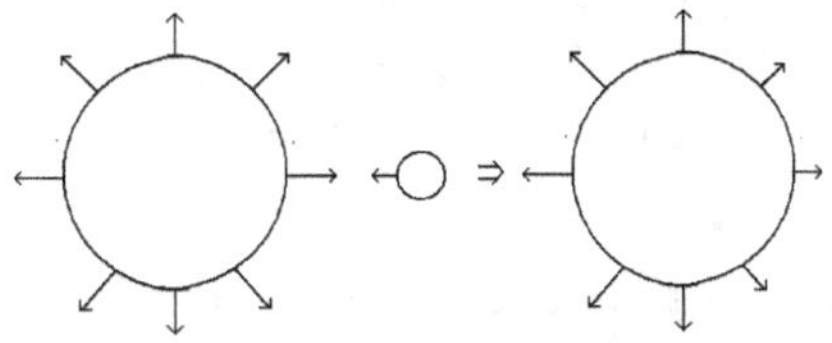

PRE-ENCOUNTER FLOW   EFFECT OF THE ENCOUNTER

Incoming Flow from a Distant Source Slows the Encountered Outward Flow
By Locally Increasing the Amounts of μ_0 and ε_0 Acting There

Figure 5-2

GRAVITATION IS THE MOMENTUM REACTION TO OUTWARD FLOW SLOWING.

The incoming *Flow* from a distant "source" particle having the effect of slowing the speed of the "encountered" particle's outward propagated *Flow* causes that "encountered" particle's outward *Flow* to have less momentum than if it were not slowed, again momentum being the product of mass and velocity.

Therefore the Newton's Third Law reaction to that reduced outward *Flow* momentum, that is the reaction back on the "encountered" particle, is smaller than otherwise. That effect takes place on the side of the "encountered particle" facing toward the "source" particle from which the slowing - causing *Flow* came.

But, on the opposite side of the "encountered" particle no such slowing of its outward propagated *Flow* is present so that the outward *Flow* there has the full natural momentum and the Newton's Third Law reaction on the particle on that side is the full natural amount. Consequently, the "encountered" particle experiencing its usual full momentum reaction back on itself on its side opposite that facing the incoming *Flow* from the "source" but experiencing reduced reaction back on itself on its side facing the incoming *Flow* from the "source", experiences a net momentum reaction toward the "source" particle from which the slowing-causing *Flow* came.

Thus the particle experiences `[1 - Cosine]` pulses of momentum increase toward the "source" gravitationally attracting particle which is gravitational acceleration.

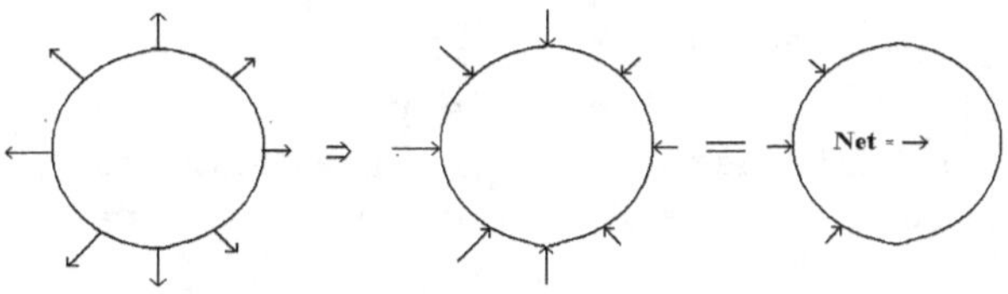

PARTIAL REDUCED FLOW CAUSES
        PARTIAL LESS REACTION AND
                NET MOMENTUM TOWARD SOURCE

Incoming Flow from a Distant Source Slows Part of the Encountered Outward Flow,
Which Reduces Part of the Reaction Momentum there,
Which Results in a Net Momentum Toward the Gravitational Source

Figure 5-3

As noted earlier the *[1 - cosine]* oscillatory form of the propagated wave is, in effect, a stream of pulses. The action of Figure 5-3 is of pulses of momentum toward the "source" gravitationally attracting particle at a repetition rate of the incoming pulses of *Flow* from that source, which is gravitational acceleration.

That is, the mass that is the "encountered" particle experiences *[1 - Cosine]* pulses of momentum increase toward the "source" gravitationally attracting particle. The "encountered" particle's mass is fixed; the momentum increases are velocity increases, an increment of velocity for each "pulse" in the stream of *Flow*. Those increments occur at the frequency of the pulses, their repetition rate as generated and propagated by the "source". That stream of velocity increments constitutes the gravitational acceleration.

DERIVATION OF NEWTON'S LAW OF GRAVITATION

In Section 4 a statement of the derived gravitational acceleration was obtained as equation *2-17*, repeated below.

(2-17)

$$\Delta v = c \cdot \frac{\delta^2}{d^2} \quad \text{per cycle of } f_{source}$$

a quite pure, precise and direct statement of the operation of gravitation. It states that gravitation is a function of the speed of light, *c*, and the inverse square law, in the context of the oscillation frequency, f_S, corresponding to the attracting, source body's mass. It should be noted that equation *4-17* is exact without involving a constant of proportionality such as *G*.

The equation *2-17* result can also be obtained directly from consideration of solely how slowing is caused by μ and ε, which demonstrates that the cause of gravitation is the slowing of wave propagation presented just above. That is as follows.

For the *Medium* of the *Propagated Outward Flow* at the instant of its propagation from its source center responding to its own μ_0 and ε_0, the value of those two are constant at what we term their free space values. Those values are inverse square reduced as the medium carrying them propagates outward from their source center-of-oscillation. (The speed of wave propagation remains the same because the waves are also inverse square reduced in amplitude.)

(5-3)   (1) At distance δ from the center of the source center,
            the first place where the propagated medium
            appears and where its concentration is greatest,
            the values of μ and ε are the free space values:

$$\mu = \mu_0 \qquad \text{and} \qquad \varepsilon = \varepsilon_0$$

          (2) Per the inverse square law, the values at distance *"d"*
            from the center of the source center are:

$$\mu(d) = \mu_0 \cdot \frac{\delta^2}{d^2} \qquad \text{and} \qquad \varepsilon(d) = \varepsilon_0 \cdot \frac{\delta^2}{d^2}$$

Then, the overall net effective values when *Flow*ing medium from a distant center passes through the outward propagation of an encountered center are

34

(5-4)
$$\mu_{net} = \left[\mu_0 + \mu_0 \cdot \frac{\delta^2}{d^2}\right] = \mu_0 \cdot \left[1 + \frac{\delta^2}{d^2}\right]$$

$$\varepsilon_{net} = \left[\varepsilon_0 + \varepsilon_0 \cdot \frac{\delta^2}{d^2}\right] = \varepsilon_0 \cdot \left[1 + \frac{\delta^2}{d^2}\right]$$

The resulting net speed of propagation is, then

(5-5)
$$c_{net} = \frac{1}{\left[\mu_{net} \cdot \varepsilon_{net}\right]^{\frac{1}{2}}} = \frac{1}{\left[1 + \frac{\delta^2}{d^2}\right] \cdot \left[\mu_0 \cdot \varepsilon_0\right]^{\frac{1}{2}}}$$

$$= \frac{c}{\left[1 + \frac{\delta^2}{d^2}\right]} = \frac{d^2}{d^2 + \delta^2} \cdot c$$

and the amount of the slowing is

(5-6)
$$\Delta c = c - c_{net}$$

$$= c \cdot \left[1 - \frac{d^2}{d^2 + \delta^2}\right]$$

$$= c \cdot \frac{\delta^2}{d^2 + \delta^2}$$

$$= c \cdot \frac{\delta^2}{d^2} \qquad \text{[d^2 is much greater than δ^2]}$$

so that

(5-7)
$$\Delta v = c \cdot \frac{\delta^2}{d^2} \qquad \text{[the slowing, Δc, equals the velocity change, Δv]}$$

which is identical to equation 2-17, above.

Equation (2-17), above, gives the gravitationally caused velocity change per cycle of the incoming gravitational wave field. The time rate of those velocity change increments, i.e. the gravitational acceleration, a_g, is Δv times the incoming wave's frequency, which is the source center's frequency, f_s.

(7-8)
$$a_g = \Delta v \cdot f_s$$

$$= c \cdot \frac{\delta^2}{d^2} \cdot f_s$$

$$= c \cdot \frac{\delta^2}{d^2} \cdot \frac{m_s \cdot c^2}{h} \qquad \text{[m_s = the source center's mass;}$$

$$= G \cdot \frac{m_s}{d^2} \qquad \text{[substituting G per equation 2-14 re the definition of the Planck Length and equation 2-16 re definition of δ]}$$

35

$(7\text{-}9)$ $F_g = a_g \cdot m_e$

$$= G \cdot \frac{m_s \cdot m_s}{d^2}$$ [m_e is the encountered center's mass.]

which is Newton's Law of Gravitation.

There is a streaming outward *Flow* of pulses of momentum from every particle and one of the effects of that stream upon its encountering another particle is to force a net acceleration of the "encountered" particle back toward the "source" particle, the effect that we call "gravitational attraction". The effect results from the "source" *Flow* combining its μ_0 and ε_0 with those of the "encountered" *Flow* causing slowing its speed of propagation.

Modification of that *Flow* before it can act on the "encountered" particle
is the means to control of gravitation.

Such modification uses the same effect as does the cause of gravitation – the μ_0 and ε_0 of particles' *Flows* combining to slow the *Flows*.

Analysis and development of the means to such modification of that *Flow* is presented in the following Sections.

37

As developed in Sections 4 and 5, there is one single streaming outward *Flow* from every particle and that *Flow* carries, produces, both the effect which we call gravitational field and that which we call electric field or, when it is changing, electro-magnetic field.

Light is an electro-magnetic field phenomenon; both light and gravitation are carried by the same streaming outward *Flow* from all particles.

Deflection of the one, light,
is deflection of the other, gravitation.

CONTROL OF GRAVITATION

Natural Deflection of Light

> There are plentiful examples of natural deflection of the *Flow* that carries light as presented below. Because that *Flow* is the sole single outward *Flow* from all particles, deflection of it is also deflection of the *Flow* that causes gravitation.
>
> By investigating deflections of light we investigate deflection of gravitation.

Light normally travels in a straight direction. But, when some effect slows a portion of the light wave front the direction of the light is deflected. In Figure 6-1 below, the shaded area propagates the arriving light at a slower velocity, v', than the original velocity, v, so that the direction of the wave front is deflected from its original direction. The index of refraction, $n = {}^c/_v$ where c is light speed and v is the velocity of propagation.

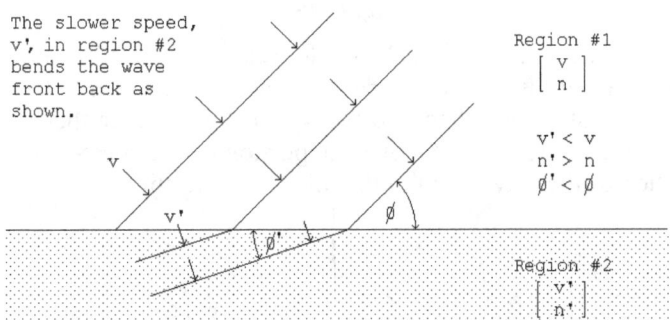

The slower speed, v', in region #2 bends the wave front back as shown.

Region #1
$\begin{bmatrix} v \\ n \end{bmatrix}$

$v' < v$
$n' > n$
$\emptyset' < \emptyset$

Region #2
$\begin{bmatrix} v' \\ n' \end{bmatrix}$

Deflection of Light's Direction by Slowing of Part of Its Wave Front
Figure 6-1

A slowing of part of its wave front is the mechanism of all bending or deflecting of light. In an optical lens, as in Figure 6-2 below, light propagates more slowly in the lens material than outside the lens. The amount of slowing in different parts of the lens depends on the thickness of the lens at each part. In the figure the light passing through the center of the lens is slowed more than that passing near the edges. The result is the curving of the light wave front.

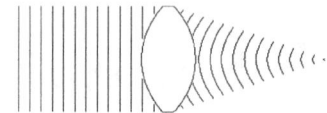

The Bending of Light's Wave Front by an Optical Lens
Figure 6-2

"Gravitational lensing", shown below, is an astronomically observed effect in which light from a cosmic object too far distant to be directly observed from Earth becomes observable because a large cosmic mass [the "lens"], located between Earth observers and that distant object, deflects the light from the distant object as if focusing it, somewhat concentrating its light toward Earth enough for it to be observed from Earth. The light rays are so bent because the lensing object slows more the portion of the wave front that is nearer to it than it slows the farther away portion of the wave front.

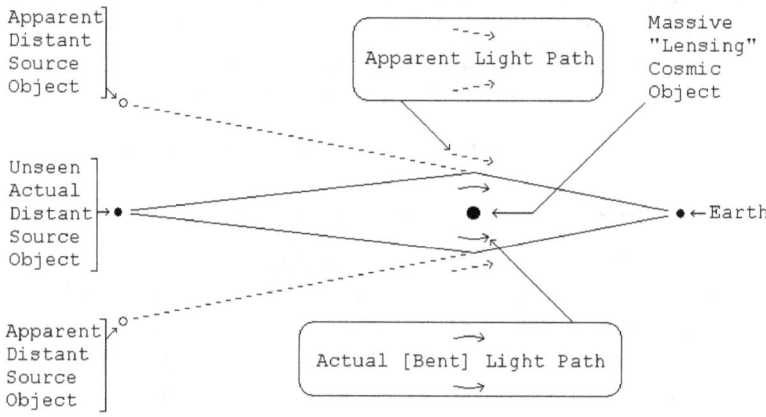

Gravitational Lensing Bending of Light Rays
Figure 6-3

The same effect occurs on a much smaller scale in the diffraction of light at the two edges of a slit cut in a flat thin piece of opaque material as shown below. The bending is greater near the edges of the slit because the slowing is greater there. The effect of the denser material in which the slit is cut slows the portion of the wave front that is nearer to it more than the portion of the wave front in the middle of the slit.

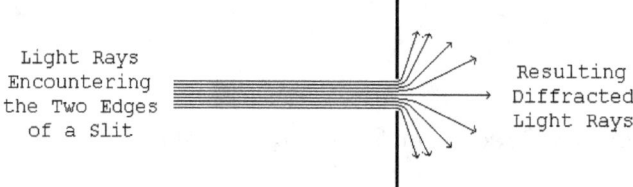

Diffraction at a Slit Causing Bending of Light Rays
Figure 6-4

In both of these cases, gravitational lensing and slit diffraction, the direction of the wave front is changed because part of the wave front is slowed relative to the rest of it. In the case of gravitational lensing the part of the wave front nearer to the "massive lensing cosmic object" is slowed more. In the case of diffraction at a slit the part of the wave front nearer to the solid, opaque material in which the slit is cut is slowed more.

But, neither of the cases, gravitational lensing and slit diffraction, involves the wave front passing from traveling through one substance to another as in the Figure 6-1 example, above. The wave front in the gravitational lensing case is traveling only through cosmic space. The wave front in the slit diffraction case is traveling only through air. There is no substance change to produce the slowing. What is it that slows part of their wave fronts thus producing the deflection ?

In the case of gravitational lensing the answer is that the effect is caused by gravitation. There is no other physical effect available. But how does gravitation produce slowing of part of the incoming wave front so as to deflect it ? Gravitation, at least as it is generally known and experienced, causes acceleration, not slowing. The answer is, of course, as follows.

As presented in the preceding Section 4, the μ_0 and ε_0 are inherent in the substance of the oscillation, which means, μ_0 and ε_0 are also inherent in the outward propagation. Each particle's *Propagated Outward Flow* contains and carries its own μ_0 and ε_0.

Upon encountering other particle's *Flow* an arriving *Flow*'s μ_0 and ε_0 (scalar quantities not vector) combine with the μ_0 and ε_0 in the "encountered" *Flow*, the combined μ_0 sum and the combined ε_0 sum being larger values than in the "encountered" original *Flow*.

The result is that that "encountered" outward *Flow* is slowed relative to its prior otherwise speed. That is, its speed of *Flow* is determined by a combination of the parameters μ_0 and ε_0 larger than its *Flow*'s otherwise natural values. The speed of *Flow* is determined by the well-established relationship:

(6-1) $$\text{Speed} = \frac{1}{\sqrt{\mu_0 \cdot \varepsilon_0}}$$

In the case of slit diffraction Figure 6-4 above, the *Flow* from the opaque material particles that are nearer to the slit is more dense, its μ_0 and ε_0 greater because of their *Flow* being less inverse square reduced, as compared to *Flow* from particles farther away. The more dense *Flow* slows the part of the light *Flow* nearer to the slit edges more than that farther from the edges.

In the case of gravitational lensing the light *Flow* nearer to the lens encounters more concentrated *Flow* from the particles of the lens, encounters greater μ_0 and ε_0.

In both cases it is the μ_0 and ε_0 in the *Flow* from the lens or slit material that adds to that in the incoming light *Flow* so as to slow the part of the light nearer to the source of the additional μ_0 and ε_0 more than that more distant.

DEFLECTING THE GRAVITATIONAL FLOW

As developed in Sections 4 and 5, there is one single streaming outward *Flow* from every particle and that *Flow* carries, produces, both the effect which we call gravitational field and that which we call electric field or, when it is changing, electro-magnetic field.

Light is an electro-magnetic field phenomenon; thus, both light and gravitation are carried by the same streaming outward *Flow* from all particles. Deflection of one, light, is deflection of the other, gravitation.

The general vertically upward outward *Flow* of gravitation can be deflected by deflecting part of a local region's gravitational *Flow* away from its normal vertical direction. Figure 6-5 below [the slit diffraction figure from earlier above but now rotated 90°] illustrates such deflection using a single slit.

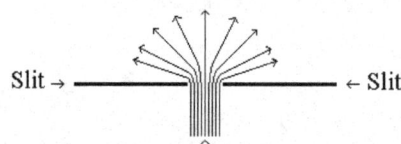

Resulting Deflected Rays of
Flow of Gravitation

Slit →   ← Slit

Rays of Flow of Gravitation
Encountering the two Edges of a Slit

Slit Diffraction, the Basic Element of a Gravitation Deflector
Figure 6-5

Multiple such slits parallel to each other would spread the deflection left and right in the figure. Additional multiple such slits at right angles to the first ones would spread the deflection over a significant area.

The edges of the slit in the above Figure 6-5 are actually rows of atoms. A cubic crystal, such as of Silicon, consists of such rows of atoms, multiple rows and rows at right angles, all equally spaced – a naturally occurring configuration of the set of slits required for deflection of gravitation.

A Small Piece of a Cubic Crystal
Figure 6-6

The *Flow* from each of the cubic crystal's atoms is radially outward. Therefore its concentration falls off as the square of distance from the atom. The amount of slowing of an incoming gravitational *Flow,* and therefore the amount of its resulting deflection, depends on the relative concentrations of the atoms' *Flow* and the overall gravitational *Flow.*

In the case of diffraction of the *Flow* of light at a slit the concentration of the *Flow* from the atoms of the slit material is comparable to the concentration in the horizontal *Flow* of the light, because the light originates from a local source, not from the Earth's immense gravitation.

But for the *Flow* from the atoms of the slit to deflect the much more concentrated vertically upward *Flow* of Earth's gravitation the *Flow* from the atoms of the slit must also be much more concentrated. The only way to achieve that more concentrated *Flow* is a configuration in which the *Flow* of Earth's gravitation is forced to pass much closer to the atoms of the slit so that, per the inverse square variation in the atoms' *Flow,* it will pass through a much greater concentration of the slit atom's *Flow*, properly designed one comparable to the concentration in the Earth's gravitational *Flow*.

The spacing between the edges of the diffracting slit is about $5 \cdot 10^{-6}$ *meters*. The spacing of the atoms at the corners of the "cubes" in a Silicon cubic crystal is $5.4 \cdot 10^{-10}$ *meters*. But, per Appendix C, the medium *Flow* concentration must be increased at least by the factor 10^{15} for its density to be comparable to Earth natural gravitation. That requires reducing the atomic spacing in the Silicon cubic crystal as follows.

(6-2) Taking Silicon's spacing of $5.4 \cdot 10^{-10}$ rounded as 10^{-11} *meters* and taking the square root of the rounded 10^{-15} as being 10^{-8} the required atomic spacing must be less than $[10^{-11}] \cdot [10^{-8}] = 10^{-19}$ *meters*.

An inter-atomic spacing of less than 10^{-19} *meters*, much closer than the natural spacing in the Silicon cubic crystal, is required to obtain deflection of a major portion of the incoming Earth's gravitational *Flow*.

Such a close atomic spacing cannot be obtained by directly arranging for, or finding a material that has, such a close atomic spacing. However, that close an atomic spacing can be effectively produced relative to just the <u>vertical</u> component of the *Flow* of gravitation by slightly tilting the Silicon cubic crystal's cubic structure relative to the vertical.

The following Figure 6-7 illustrates the tilting, schematically not to scale, and shows how it increases the number of crystal atoms closely encountered by the upward gravitational *Flow*.

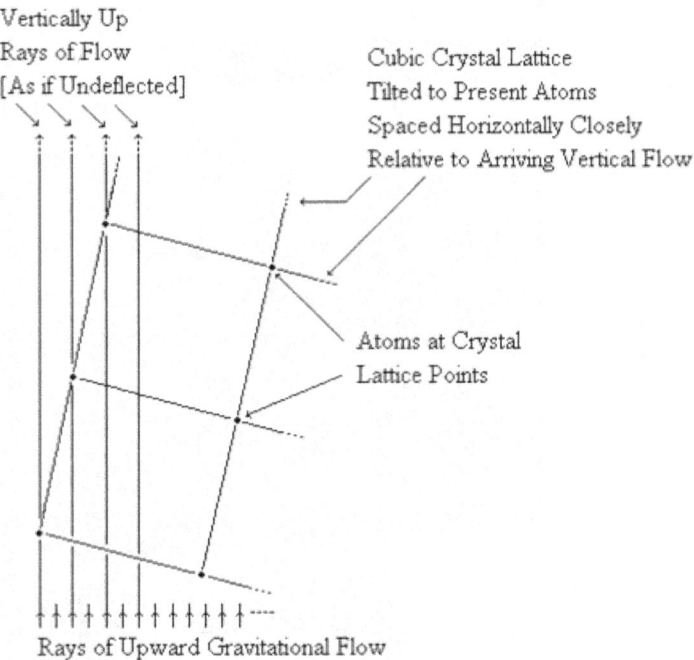

Cubic Crystal Lattice Tilted for Effective Gravitational Flow Deflection
Figure 6-7

By appropriate tilting of the cubic structure each of its $5.4 \cdot 10^{-10}$ *meters* inter-atomic spaces can be effectively sub-divided into 10^{10} "sub-spaces" each of them $5.4 \cdot 10^{-20}$ *meters* long and with an atom in each. A 4.5 *mm* shim on a 30 *cm* diameter Silicon cubic crystal ingot would produce such an effect, producing a tilt *tangent = 0.015* for a *tilt angle = 0.86°* that would produce the objective effective sub-division of the crystals' natural inter-atomic spacing, a sub-division that acts only on vertical *Flow* as components of gravitation.

Pure, monolithic, Silicon cubic crystals up to 30 *cm* in diameter are grown for making the "chips" used in many electronic devices. The gravitation deflector requires, instead of the thin wafers sawed from the "mother" crystal for "chip" manufacture, a large, thick ingot of Silicon cubic crystal a half meter or more in length as is the "un-sawed" mother crystal.

However, the foregoing analysis only presents the general principles on which a gravitation deflector could be based. We now proceed to detailed mathematical analysis of the problem and of the solution.

Analysis of the Amount of Deflection

The *Propagated Outward Flow* concentration produced by an atom of the slit edge falls off inversely as the square of distance from it.

The Cauchy-Lorentz Distribution is an inverse square function that can represent the diffraction pattern envelope. <u>The envelope of the pattern is the relative amounts of the underlying *Propagated Outward Flow* carrying the light</u>.

The relative amounts of the incoming *Propagated Outward Flow* that are deflected in any specific direction can thus be calculated with the Cauchy-Lorentz distribution.

THE AMOUNT OF DEFLECTION

The manner of the deflection is curving of the path of rays of gravitational Flow as they pass close to atoms of the deflector with the direction to which curved depending on the relative positions of the ray and an atom and the amount of the curving depending on how close the ray passes to the atom. Because of the range of those variables and their various combinations the "deflection" is essentially a "scattering" in various amounts in various directions, all scattering being away from the perfectly vertical upward which the deflector is designed to solely deflect. The "scattering" is illustrated two-dimensionally in the figure below visualized as that figure viewed from the top and rotated through a full circle.

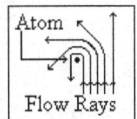

Single Atom Deflection of Rays of Gravitational Flow

A two-dimensional physical example of the "scattering" is the diffraction pattern of light diffracted by a slit. Figure 7-1, below, presents the diffraction pattern produced by a slit that is $5.4 \cdot 10^{-6}$ meter wide with incoming light of wavelength $4.13 \cdot 10^{-7}$ meter. The peaks and valleys of the pattern, the interference pattern, are a phenomenon of the light imprint on the *Flow* that carries it. The envelope of the pattern is the relative amounts of the underlying *Flow* carrying the light.

For that reason, while the interference pattern varies according to the slit width and the wavelength of the light involved, the form of the envelope of that pattern is always the same.

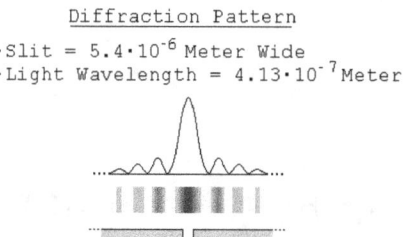

Diffraction Pattern
·Slit = $5.4 \cdot 10^{-6}$ Meter Wide
·Light Wavelength = $4.13 \cdot 10^{-7}$ Meter

Figure 7-1 - A Slit Light Diffraction Pattern

The *Flow* concentration produced by the two slit edges falls off with distance from the slit edge inversely as the square of distance from its atoms. The Cauchy-Lorentz Distribution is an inverse square function of its variable. Its Density Function can represent the relative *Flow* intensity pattern produced by the diffraction process by representing the envelope of the diffraction pattern. In Figure 7-2, below, the Cauchy-Lorentz distribution is fitted to the diffraction pattern by the appropriate choice of value of its distribution parameter γ [Greek *gamma*].

The Envelope of the Relative Intensities of the Light Diffraction Pattern Is the Actual Amount of the *Flow* Relative Intensities.

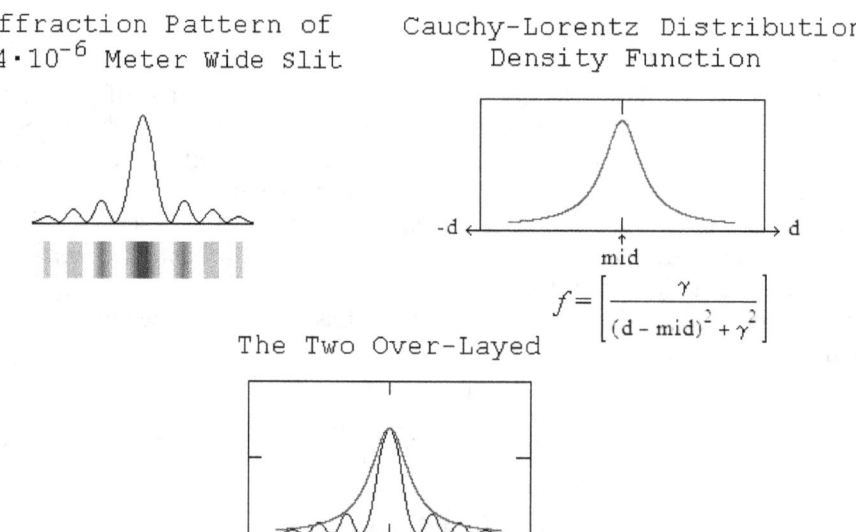

Diffraction Pattern of $5.4 \cdot 10^{-6}$ Meter Wide Slit

Cauchy-Lorentz Distribution Density Function

$$f = \left[\frac{\gamma}{(d - mid)^2 + \gamma^2} \right]$$

The Two Over-Layed

Figure 7-2 – The Cauchy-Lorentz Distribution Diffraction Pattern Envelope

The deflection angle, Φ *[Greek Phi]*, is the angle of deflection of the rays to any particular point on the diffraction pattern. That is Φ is the angle of deflection of the rays directed to that particular point and of intensity per the Cauchy-Lorentz Distribution at that point.

The interest here is not in the location of the light interference maxima and minima, but in the deflection angles the diffraction imposes on the *Flow*. However, calculation of the deflection angles to the minima provides a good indication of the amount of *Flow* deflection obtained over the overall diffraction pattern. The table below presents that data for the $5.4 \cdot 10^{-6}$ meter wide slit with incoming light of wavelength $4.13 \cdot 10^{-7}$ meter. [The minimums are counted outward from the center peak of the diffraction interference pattern].

Minimum #	$\Phi°$		Minimum #	$\Phi°$
1	4.39		8	37.72
2	8.80		9	43.50
3	13.26		10	49.89
4	17.81		11	57.28
5	22.48		12	66.60
6	27.36		13	83.86
7	32.37		14	Sin(Φ) > 1.0

$$Sin(\Phi) = n \cdot [\, light\ wavelength\, /\, slit\ width\,],\ n = 1, 2, ...$$
Figure 7-3 – Table of Diffraction Minimums Deflection Angles

Again, while we are not interested in the diffraction minimums and not in the diffraction interference patterns at all, the envelope of the diffraction pattern depicts the distribution of the deflection of the *Propagated Outward Flow* that carried the light in the diffraction pattern.

The above table demonstrates that the deflection of the *Flow* is at least in amounts up to $90°$. That deflection may well extend to angles beyond $90°$, perhaps to as much as $180°$, a complete reversal of direction. There is no way of determining that from the diffraction pattern, however, because the light of the diffraction pattern cannot be deflected beyond $90°$ because the light cannot penetrate the material containing the slit.

But, the *Flow* readily penetrates and permeates all of material reality.

The tilt of the cubic crystal structure divides the slit into a large number of sub regions the first and last of which are at the slit's edge and produce the maximum deflection. The tilt also arranges that ultimately all of the vertical components of the incoming vertical Flow must pass through one of those "at the edge of the slit" regions, *i.e.* must experience maximum deflection.

The overall average effect is equivalent to every ray's vertical component curving at least *90°* because the crystal tilt causes every ray to pass extremely close to an atom at some point in the crystal, as shown for the extreme rays in the figure below.

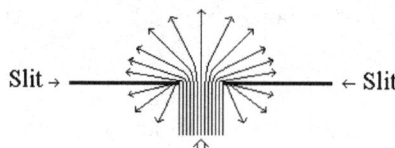

Resulting Deflected Rays of
Flow of Gravitation

Slit → ———————— ← Slit

Rays of *Flow* of Gravitation
Encountering the two Edges of a Slit

Figure 7-4 – Single Slit Gravitation Deflection

PROPAGATED OUTWARD FLOW DEFLECTION CAUSED BY ITS FLOW SLOWING

The bending of *Propagated Outward Flows'* paths results from differential slowing, that is the systematic slowing of the *Propagated Outward Flow* wave front in different amounts along that front. The slowing takes place in accordance with equation *(5-2)*. Figure 7-5, below, depicts the differential slowing-caused process.

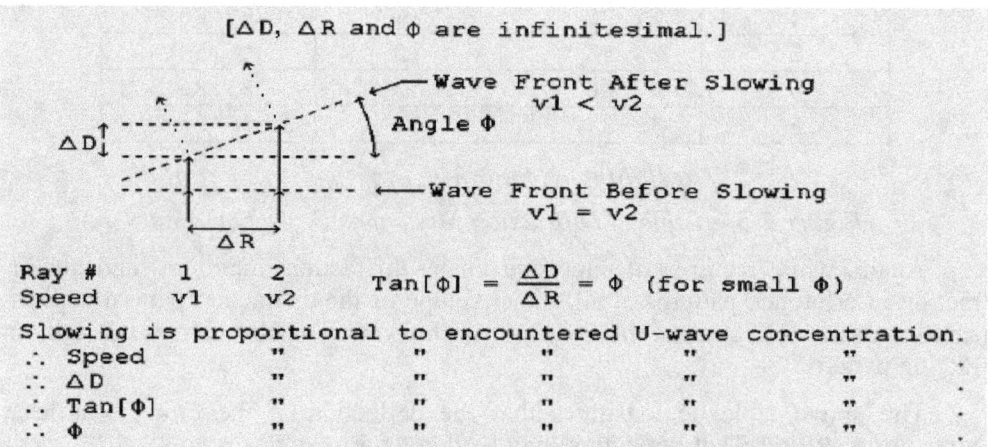

Figure 7-5 – Propagated Outward Flow Deflection

The figure indicates the differential slowing of the upward-directed [as for gravitation] *Propagated Outward Flow* flux that results in deflection of the *Propagated Outward Flows'* paths. The slowing is directly proportional to the encountered concentration of the *Propagated Outward Flow* flux, and, therefore the angle of deflection, Φ, is proportional to that concentration.

QUANTIFYING THE PROPAGATED OUTWARD FLOW DEFLECTION IN LIGHT

The diffraction pattern is a projection on a screen or piece of photographic film of the diffracted light as it spreads out due to the diffracting action. The physical size, the linear dimension of the pattern becomes larger as the distance from the diffracting slit to the screen or film on which the pattern appears increases. But the angles, as measured from the center of the slit to any point on the diffraction pattern [relative to the 0° angle from the center of the slit to the center of the pattern], are the same regardless of the distance from the slit to the screen or film.

50

Therefore, to analyze and evaluate the pattern requires attending to those angles, not linear distances on the pattern. Since the linear distances on the pattern are irrelevant, any convenient distance from the slit to the screen or film may be chosen. In the following analysis that distance will be taken as equal to the slit width, $5.4 \cdot 10^{-6}$ *meter* in this case.

The data of interest here, which is a measure of the amount of *Propagated Outward Flow* bending contained in the diffraction pattern, is the portion of the total light incident on the slit appearing in any specified portion of the diffraction pattern. That portion can be defined in terms of the angles just described and that portion is an otherwise dimensionless number, again independent of the physical or linear size of the diffraction pattern.

The Cauchy-Lorentz Distribution for this application is as follows.

(7-1) The Cauchy-Lorentz Distribution Density Function

   [a] Underline General

$$ f(x;x_0,\gamma) = \frac{1}{\pi} \cdot \left[\frac{\gamma}{(x - x_0)^2 + \gamma^2} \right] $$

   [b] Underline As Used Here

$$ f(d;mid,\gamma) = \left[\frac{\gamma}{(d - mid)^2 + \gamma^2} \right] $$

   mid = half-way point between slit edges
   d = distance from mid
   γ = half-width at half-maximum

From the above Figure 7-2, the half-width of the Cauchy-Lorentz Distribution at its half-maximum is 74.0% of the distance from the mid-point to the first minimum in the interference pattern. That is γ is 74.0% of the displacement from the centerline to the first intensity minimum outward from the centerline. Calculating the deflection angle to that minimum the angle is found to be $4.39°$.

The corresponding displacement along the d-axis [for screen distance = slit width] of Figure 7-3 is the value of γ in the Cauchy-Lorentz distribution.

(7-2) γ = [74% of] [[slit width] $\cdot$ Tan[4.39°]]

   = $[0.74] \cdot [5.4 \cdot 10^{-6}$ meter] $\cdot [0.077]$

   = $3.1 \cdot 10^{-7}$ meter

The deflection angle, Φ, for any particular point on the diffraction pattern is the angle between [a] a reference line that runs from the center of the slit perpendicular to the barrier containing the slit toward the projected diffraction pattern and [b] a line running from the center of the slit to the location of the particular point on the diffraction pattern. That is the angle of deflection of the rays directed to that point and of intensity per the Cauchy-Lorentz Distribution at that point.

In these diffraction patterns so long as the ratio of the wavelength of the incident light to the width of the slit is constant, then each deflection angle, Φ, is independent of the distance from the slit to the screen where the diffraction pattern is projected.

The Cauchy-Lorentz Distribution's Cumulative Distribution Function is the integral of the Density Function, that is the area under the Density Function curve, the cumulative density. That function is given in equation *(7-3)*, below.

(7-3) The Cauchy-Lorentz Distribution Cumulative
Distribution Function

[a] In General

$$f_{cum}(x; x_0, \gamma) = \frac{1}{\pi} \cdot arctan\left[\frac{x - x_0}{\gamma}\right] + \frac{1}{2}$$

[b] As Used Here

$$f_{cum}(d; mid, \gamma) = \frac{1}{\pi} \cdot arctan\left[\frac{d - mid}{\gamma}\right] + \frac{1}{2}$$

With *mid = 0*, when *d = -∞* [a deflection of *90°* to the left in Figure 7-3], then $f_{cum} = 0$. Likewise at *d = +∞* then $f_{cum} = 1$, the total amount. To find the fraction, *F*, of the total amount of the incident light entering the slit that is deflected through some chosen angle, *Φ*, or more to the left of *mid* the procedure is as follows, taking *Φ = -45°* as an example and using $\gamma = 3.1 \cdot 10^{-7}$ meter per equation *(7-2)*. Because that light exists only on the *Propagated Outward Flows* carrying it the portion, *F*, is the fraction of the total amount of *Propagated Outward Flows* entering the slit that is deflected through angle *Φ* or more.

1 – Calculate the displacement, *d,* of Figure 7-3.

(7-4) $d = Tan[\theta] \times [slit\ width]$

$= Tan[-45°] \times [5.4 \cdot 10^{-6}]$

$= -5.4 \cdot 10^{-6}$ [for this example of $\theta = -45°$]

2 – Calculate $F = f_{cum}(d; mid, \gamma)$ from equation *(7-3)*.

(7-5)
$$F = f_{cum}(d; mid, \gamma) = \frac{1}{\pi} \cdot arctan\left[\frac{d - mid}{\gamma}\right] + \frac{1}{2}$$

$$= \frac{1}{\pi} \cdot arctan\left[\frac{(-5.4 \cdot 10^{-6}) - (0)}{3.1 \cdot 10^{-7}}\right] + \frac{1}{2}$$

$$= 0.018$$

Then P, the percentage deflected through angle *Φ* or more of the total *Propagated Outward Flows* incident on the slit is:

$F \div f_{cum}(d = +\infty) = F \div 1 = F.$

P = 1.8% of total incident light entering the
slit on each side [for this example].

In this example calculation the portion of the total *Propagated Outward Flow* flux that is deflected by *Φ = 45° or more* is $P_{45} = 1.8 + 1.8 = 3.6\%$.

Table 7-6, below, presents the portion of the total amount of the incoming gravitational *Propagated Outward Flow* flux that is deflected through some chosen angle, *Φ* or more, using the above *45°* example type of calculations for each of the deflection angles cited in Table 7-3, above.

$\Phi°$	% Deflected	$\Phi°$	% Deflected
4.39	40.9	37.72	4.7
8.80	22.6	43.50	3.8
13.26	15.2	49.89	3.1
17.81	11.3	57.28	2.3
22.48	8.8	66.60	1.6
27.36	7.1	83.86	0.4
32.37	5.7		

Table 7-6
*Percent of Total Propagated Outward Flow that is Deflected By Various Angles of
Deflection, Φ, or More*

USING THESE SLIT DIFFRACTION RESULTS FOR
A GRAVITATION DEFLECTOR

The above table and example indicate that significant *Propagated Outward Flow* ray deflection does take place in the case of the atoms along the edge of the $5.4·10^{-6}$ *meter* wide slit, but the amount of deflection is not very much – about only *3.6%* deflected *45°* or more, in the example.

On the other hand, looking at *100%* of the rays of *Propagated Outward Flow* flux that arrive, uniformly spaced, at the $5.4·10^{-6}$ *meter* wide slit, *3.6%* of them arrived at that slit near enough to the atoms of one of the edges so as to be deflected *45°* or more. All of the rays of that *3.6%* achieved that much deflection because they passed their deflecting atom much more closely than the rest of the rays.

The *1.8%* on each side of the Cauchy-Lorentz Distribution passed its deflecting atom within a distance of *1.8%* of the slit width $[0.018 × (5.4·10^{-6}) = 9.7·10^{-8}$ *meter]*. If it could be arranged that all of the vertically upward *Propagated Outward Flow* gravitational flux were to pass that closely to atom then *100%* of the gravitational flux would be deflected by *45°* or more.

However, these deflection calculations are for a *Propagated Outward Flow* flux of the density or concentration of the *Propagated Outward Flow* carrying the beam of light to the diffracting slit. The vertically upward *Propagated Outward Flow* flux of the Earth's gravitational field is immensely more dense and concentrated.

The following Section 8 addresses that aspect.

53

Gravitation Deflector Mathematical Analysis Details

> The Cauchy-Lorentz light-and-slit analysis deflections and calculations in Section 7 were for light traveling in the *Propagated Outward Flow* flux density generated by an Earth surface light source not the much more concentrated Earth overall gravitational *Propagated Outward Flow* flux concentration.
> As developed in Appendix C, the ratio of the two, Earth gravitation and local light, is on the order of 10^{15}.
> The calculations of Section 7 must now be adjusted for that ratio.

A Cubic Crystal Deflector

A small portion of a Silicon cubic crystal is depicted below.

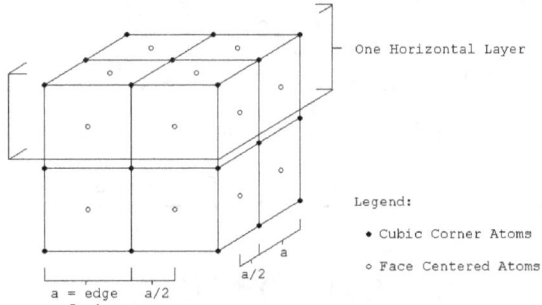

Figure 8-4
A Silicon Cubic Crystal

In the Silicon cubic crystal the edge of the cube, a, is $5.4 \cdot 10^{-10}$ *meters*. The effective horizontal interatomic spacing for vertically upward traveling *Propagated Outward Flow*s is half the edge, $a/2 = 2.7 \cdot 10^{-10}$ *meters*. From the figure the vertical layer thickness is $a = 5.4 \cdot 10^{-10}$.

The edges of the slit that produces the light diffraction pattern of Figure 7-2 consist of atoms spaced along the slit edge at an interatomic spacing that is essentially the same as a cubic crystal's interatomic spacing, of about $2.7 \cdot 10^{-10}$ *meter*. The only difference between the light diffraction $5.4 \cdot 10^{-6}$ *meter* wide slit and a cubic crystal's interatomic spacing is that in the cubic crystal the "slit" width is that same interatomic spacing, about $2.7 \cdot 10^{-10}$ *meter*.

The diffraction pattern of Figure 7-2 is determined by the edges of the slit. The edges are the limit of the "slice" of incident light that passes through the slit and the light at those edges is the most deflected because it is the nearest to the deflecting atoms of the slit edge. Similarly, the edges jointly define the mid point of the diffraction pattern which is where the action of the two edges are equally strong so that their deflecting effects cancel each other to no net deflection.

In the cubic crystal those defining points are only apart $2.7 \cdot 10^{-10}$ *meter* as compared to $5.4 \cdot 10^{-6}$ *meter* apart in the case of the slit. The calculations of equations 7-1 through 7-5 must be re-calculated for that $2.7 \cdot 10^{-10}$ *meter* slit. That requires evaluating γ for its Cauchy-Lorentz Distribution. That is the same as in equation 7-2 except that the value of the slit width is changed to $2.7 \cdot 10^{-10}$ *meter*. The result is equation *(7-2')*.

(7-2') γ' = [74% of] [[slit width]·Tan[4.39°]]

$\qquad$ = [0.74]·[$2.7 \cdot 10^{-10}$ *meter*]·[0.077]

$\qquad$ = $1.5 \cdot 10^{-11}$ *meter*

Calculating the portion, P, of the total amount of the incident *Propagated Outward Flow*s entering that slit that is deflected through $\theta = -45°$ to the left of the mid point of the diffraction pattern and its Cauchy-Lorentz Distribution using $\gamma = 1.5 \cdot 10^{-11}$ *meter* per equation *(3-2')* is as follows.

$\qquad$ 1 – Calculate the displacement, d.

(3-4') $\quad$ d = Tan[θ] × [slit width]

$\qquad$ = Tan[-45°] × [$2.7 \cdot 10^{-10}$]

$\qquad$ = $-2.7 \cdot 10^{-10}$ [this example of θ = -45°]

$\qquad$ 2 – Calculate $P = f_{cum}(d;mid,\gamma)$ from equation (3-3).

(3-5')
$$P = f_{cum}(d;mid,\gamma) = \frac{1}{\pi} \cdot \arctan\left[\frac{d - mid}{\gamma}\right] + \frac{1}{2}$$

$$= \frac{1}{\pi} \cdot \arctan\left[\frac{(-2.7 \cdot 10^{-10}) - (0)}{1.5 \cdot 10^{-11}}\right] + \frac{1}{2}$$

$$= 0.018$$

Again the portion of the total *Propagated Outward Flow* flux that is deflected by $\theta = 45°$ or more is $P_{45} = 1.8\% + 1.8\% = 3.6\%$. The result is unchanged from that in the case of the $5.4 \cdot 10^{-6}$ *meter* slit. The reason for that is that the parameters of the Cauchy-Lorentz Distribution describing the deflected *Propagated Outward Flow* amounts in the various directions of deflection are determined by the two opposed slit edges. Contracting their spacing correspondingly contracts the distribution.

Now for the 1.8% on each side of the Cauchy-Lorentz Distribution to pass its deflecting atom within a distance equal to 1.8% of the slit width, the new value of that distance is the value for the cubic crystal slit, $[0.018 × (2.7 \cdot 10^{-10}) = 4.9 \cdot 10^{-12}$ *meter]*. If it could be arranged that all of the vertically upward *Propagated Outward Flow* gravitational flux were to pass within that close a distance of an atom of the cubic crystal lattice, then 100% of the gravitational flux should be deflected by $45°$ or more.

EARTH'S GRAVITATION VS. A SURFACE LIGHT SOURCE

However the light-and-slit analysis deflections and calculations in Section 7 were for light traveling in the *Propagated Outward Flow* flux density generated by the Earth surface light source not the much more concentrated Earth overall gravitational *Propagated Outward Flow* outward flux. The deflections and calculations for diffraction of light as developed in Section 7 must be adjusted to compete at the level of Earth gravitational *Propagated Outward Flow* flux rather than at that of an Earth surface light source if there is to be a noticeable deflecting affect on Earth gravitation.

The *Propagated Outward Flow* flux of interest, that is that involved in the light-and-slit diffraction effects, is that flowing in the same direction as the beam of light that is diffracted at the slit. That flux derives partly from the ambient air [the flux if the light source were removed] and partly from the light source. At the slit the two are inter-mixed and the diffracting action deflects all *Propagated Outward Flows* indiscriminately, not the light *Propagated Outward Flows* selectively.

The ratio of the Earth's surface gravitational acceleration, $9.8 \ ^m/_{sec2}$, to, from Table C-8, the gravitational acceleration of air, $4.81 \times 10^{-17} \ ^m/_{sec2}$, is about $2 \cdot 10^{17}$. From that table, the gravitational acceleration of metals is on the order of $10^{-14} \ ^m/_{sec2}$ as compared to the Earth's overall gravitational acceleration of about $9.8 \ ^m/_{sec2}$ for a ratio of about 10^{15}. Consequently, the flux actually carrying the light [generated by a metallic light source] and entering the slit is the dominant factor.

The *Propagated Outward Flow* fluxes are proportional to the acceleration that they produce. The ratio of the accelerations, which is the ratio of the *Propagated Outward Flow* fluxes, as developed in Appendix C, is as given in equation 8-1, below.

$$(8\text{-}1) \quad \text{Ratio} = \frac{\text{Acceleration of Earth Gravity}}{\text{Acceleration of Diffracted Light } Flow}$$

$$= \frac{\text{Earth Gravity } Propagated \ Outward \ Flows \text{ Flux}}{\text{Slit Diffracted Light } Propagated \ Outward \ Flows \text{ Flux}}$$

$$\approx 10^{15}$$

Therefore the *Propagated Outward Flow* concentration which all vertical rays of gravitational *Propagated Outward Flow* flux must be forced to encounter by being forced to pass close to the cubic crystal's atoms must for this purpose be made 10^{15} times greater. The gravitational *Propagated Outward Flow* flux must be forced to pass accordingly even closer to the cubic crystal's atoms.

However, the *Propagated Outward Flow* concentration from the atoms is inverse-square reduced with distance from the atom and accordingly so increases with nearness to the atom. Consequently, to increase the concentration by a factor of 10^{15} requires reducing the separation distance by a factor of only the square root of that, about $3.2 \cdot 10^7$.

The earlier above found effective interatomic spacing to be forced by tilting the cubic crystal, $2 \times [4.9 \cdot 10^{-12}] = 9.8 \cdot 10^{-12} \ meter$, must now be that divided by $3.2 \cdot 10^7$ the result for which is $3 \cdot 10^{-19} \ meter$. That arrangement, arranging that all of the *Propagated Outward Flow* gravitational flux must, at some layer, pass within $3 \cdot 10^{-19} \ meter$ of an atom of the cubic crystal will result in essentially 100% of the gravitational *Propagated Outward Flow* flux passing so close to some atom that it should be deflected by $45°$ or more..

With the cubic crystal's natural interatomic spacing being $2.7 \cdot 10^{-10}$ $meter$ and the effective spacing to be forced is $3 \cdot 10^{-19}$ $meter$ then each natural interatomic space must be sub-divided into $9 \cdot 10^8$ "pieces". If the crystal is tilted such that each of the layers of the crystal lattice is located offset from the layer below it by $[^1/_{9 \cdot 10^8}] \cdot [2.7 \cdot 10^{-10}]$ $= 3 \cdot 10^{-19}$ $meter$ in each of the two horizontal directions of the orientation of the lattice then the objective is met.

The direct implementation of that would require a tilt at an angle whose tangent is the offset divided by the interatomic [layer-to-layer] spacing, that is $[3 \cdot 10^{-19}] \div [5.4 \cdot 10^{-10}] = 5.6 \cdot 10^{-10}$, an angle of about $6.4 \cdot 10^{-8}°$. That means that the tilt causes each successive layer to offer its atoms a further $3 \cdot 10^{-19}$ $meter$ offset so that enough layers will produce offering the atoms at every $3 \cdot 10^{-19}$ $meter$ increment in each $2.7 \cdot 10^{-10}$ $meter$ horizontal interatomic space. See Figure 8-5, next page.

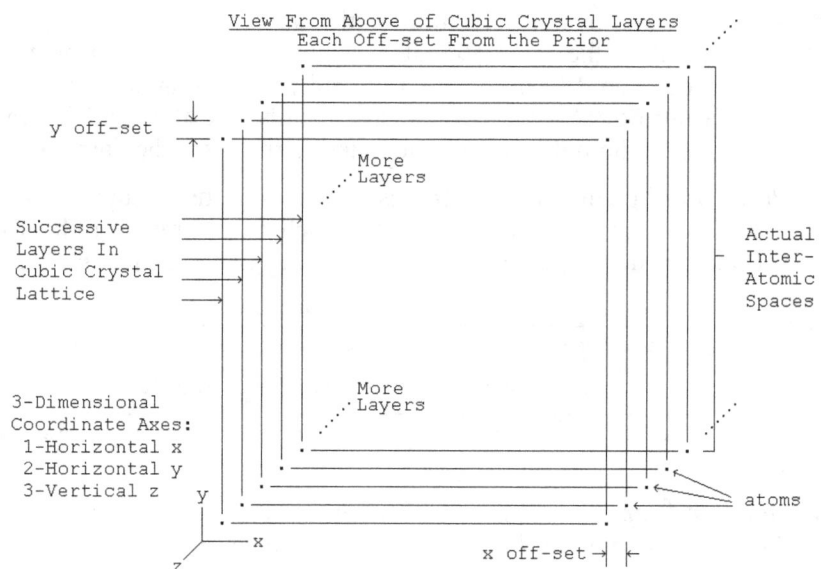

Figure 8-5
Crystal Layers, Offset Slightly, Achieving Effective
Close Interatomic Spacing
[The z-axis is vertical. The x- and y-axes are in layers.]
[Not to Scale.]

The required number of layers is one layer for each of the $9 \cdot 10^8$ "pieces" into which each $2.7 \cdot 10^{-10}$ $meter$ horizontal interatomic space is divided: $9 \cdot 10^8$ layers.

Such a fine tilt angle and its precision are unlikely to be able to set up. The solution to that is that the successive layers need not each supply the minute offset relative to their adjacent layers. If the layers as depicted in Figure 8-5, above, were shuffled into any order whatsoever, they would still have the same effect that no vertical ray could avoid passing within $3 \cdot 10^{-19}$ $meter$ of an atom, some atom, not necessarily one in the immediately next layer.

Of course, the layers in the cubic crystal cannot be shuffled or re-arranged, but that is not necessary. All that is necessary to operate using a larger tilt angle is that the same sufficient number of layers overall be employed and that the tilt be such ["workable tilt"] that the actual *x-axis offset* and the actual *y-axis offset* be such that, after that "same sufficient number of layers overall", each required effective atomic position appears somewhere, in some layer, even though not necessarily in "sequential order".

"Unworkable tilts" are those that duplicate needed atomic positions or that fail to produce all needed atomic positions, or both. The problem of what successfully workable tilts are and how they relate to unworkable tilts is developed in Appendix D, Factors Affecting Cubic Crystal Tilt.

The number of layers required, $9 \cdot 10^8$, requires a cubic crystal thickness of that number of layers multiplied by the individual layer thickness, which is $9 \cdot 10^8 \times 5.4 \cdot 10^{-10} = 0.50$ *meters* or 50 *cm.*

Each "slit" in the cubic crystal is a pair of atoms spaced apart horizontally by the crystal lattice interatomic spacing of $2.7 \cdot 10^{-10}$ *meter*; or, more precisely, each "slit" is a linear "string" of such atom pairs, in any single layer of the crystal, and running from one side to the other of the crystal just as the slit edges in the case of light diffraction by a slit is a linear "string" of the atoms of which the slit edge consists.

For each such $2.7 \cdot 10^{-10}$ *meter* wide "slit" the above tilt procedure implements arranging that out of all of that portion of the total gravitational *Propagated Outward Flow* flux that passes through it, the $^1/_9 \cdot 10^8$ or $1.1 \cdot 10^{-7}$% on either side, a total of $2.2 \cdot 10^{-7}$%, passes within $3 \cdot 10^{-19}$ *meter* of an atom of the cubic crystal lattice and will be deflected by $45°$ or more away from its pre-deflection vertically upward direction. That leaves the issue of what happens to the balance of the gravitational *Propagated Outward Flow* flux entering each such "slit".

The *Propagated Outward Flow* propagation of each such atom falls off in concentration inversely as the square of the distance from it. The $3 \cdot 10^{-19}$ *meter* closeness is required to obtain the $45°$ deflection. Of the total gravitational *Propagated Outward Flow* flux entering that slit, at ten times farther away from an atom, $3 \cdot 10^{-18}$ *meter*, the concentration is reduced by a factor of $[^1/_{10}]^2 = {}^1/_{100}$. There the angle of deflection is reduced by approximately that factor to about $0.45°$. That deflection is experienced by about $1.1 \cdot 10^{-6}$% on either side out of the total gravitational *Propagated Outward Flow* flux that passes through the "slit", a total of $2.2 \cdot 10^{-6}$%.

Still farther away, at $3 \cdot 10^{-17}$ *meter* from an atom, the concentration is reduced by a factor of $[^1/_{100}]^2 = {}^1/_{10,000}$. There the angle of deflection is reduced by approximately that factor to about $0.0045°$ and applies to about $2.2 \cdot 10^{-5}$%.

Thus far more than 99 % of the total *Propagated Outward Flow* flux entering the "slit" experiences negligible deflection. That is, until layer-by-layer in the crystal lattice further portions having earlier experienced that negligible deflection then experience the "$3 \cdot 10^{-19}$ *meter*" condition until, eventually, all of the *Propagated Outward Flow* flux experiences the "$3 \cdot 10^{-19}$ *meter*" condition and is deflected by $45°$ or more.

Examining a mean free path analysis it is now found to be the case that the target interatomic spacing to be achieved by the tilt of the cubic crystal is $3 \cdot 10^{-19}$ *meters*. The mean free path in the Earth for that same $3 \cdot 10^{-19}$ *meters* target size then is calculated as follows.

$(8-2)$ $\text{MFP} = {}^1\!/_{\text{C} \cdot \text{A}}$

$$= \frac{1}{[\text{C, Atoms Per Unit Volume}] \times [\text{A, Atom Cross Section Area}]}$$

For the Earth the concentration of atoms is on the order of $C = 5 \cdot 10^{28}$ *per cubic meter*. In the cubic crystal deflector the target spacing achieved by the tilt is $3 \cdot 10^{-19}$ meters. Each target has cross sectional area space available to it equal to a circle of that diameter so that

$(8-3)$ $A = {}^\pi\!/_4 \cdot [3 \cdot 10^{-19}]^2 = 7.1 \cdot 10^{-38}$ *meter^2*

and, for such targets the mean free path in the Earth's outer layers is

$(8-4)$ $\text{MFP} = 2.8 \cdot 10^8$ *meters*.

That is to be compared to the mean free path in the cubic crystal deflector being one-half the cubic crystal thickness of 0.50 *meters* or 0.25 *meters*.

The gravitation deflector is about 10^{10} times more effective than the natural Earth at intercepting Earth's natural gravitation.

However, that effectiveness is only for vertical rays of *Flow*.

The Silicon crystal's mean free path for non-vertical *Flow* – *Flow* already once deflected within the crystal and any non-vertical component of all of the gravitational *Flow* – is that of Earth, $2.5 \cdot 10^9$ *meters*, which takes such *Flow* out of the crystal.

Other Factors Affecting Cubic Crystal Tilt

Appendix D treats various other factors affecting cubic crystal tilt such as temperature, thermal vibrations and black body radiation and the solution to dealing with them.

The discussion can now proceed from theory to engineering design of gravitation deflectors.

Anti – Gravitational Acceleration

With the material universe exhibiting its many instances of balances of opposites an anti-gravitational effect is to be expected.

THE DEFLECTION CAUSES A REACTION BACK ON THE DEFLECTOR

Everything in nature is balanced. Nature exhibits a general law of conservation that goes far beyond conservation of energy. For example:

- All positive charge is ultimately, somewhere, balanced by an equal amount of negative charge;

- Gravitational attraction takes place by a mass acting on another mass. The attractive force acting on each is the same in magnitude and opposite in direction; the forces balance;

- The "Big Bang" produced equal amounts of matter and anti-matter;

- Newton's Third Law: For every force there is an equal-but-opposite reaction force;

- Every North magnetic pole is matched by an equal strength South magnetic pole.

As with that balance, there is a reaction on the deflection-causing gravitation deflector, a reaction to its deflecting action, a balancing reaction.

DESCRIPTION OF THE ANTI-GRAVITATIONAL EFFECT

The gravitational field *Propagated Outward Flow* is an essentially unlimited capacity to produce acceleration. That capacity is what the outward propagating gravitational field *Propagated Outward Flow* continuously does; it accelerates any and every encountered particle of mass no matter how great or how many and no matter where located. But, the amount of gravitational acceleration does not depend on the mass of the encountered particle that is accelerated; rather, it is in an amount dependent only on the mass, M, of the gravitational *Flow* source and the distance, d, from that source to the accelerated mass, which two parameters determine the gravitational field strength at the accelerated mass.

```
Gravitational Acceleration =  G ·M/d2
```

That *Flow* is what the gravitational deflector deflects.

63

The associated "force" is that acceleration multiplied by the mass that is accelerated, which can be whatever mass it happens to be. Thus for gravitation the "force" is inconsequential. No "force" is actually there except in our mental concept of the action. It is the acceleration that is the action.

The reaction on the deflector is an "equal but opposite" <u>acceleration of the deflector mechanism away from the source of the gravitational field *Flow*</u>; that is, it acts in the opposite direction from the direction, toward the source, of the acceleration that undeflected gravitation produces. The deflector experiences that reaction acceleration regardless of the mass of the deflector and no matter what additional mass may be attached to it, which attached mass is accelerated with the deflector.

That is because, again, gravitational field *Flow* accelerates any and every encountered particle of mass no matter how many and no matter where located, in amount independent of the mass accelerated, the amount dependent only on the gravitational field strength at the encountered mass.

$$\text{Gravitational Acceleration} = G \cdot M / d^2$$

The direction of the reaction-produced acceleration [repulsion] is the opposite of the direction [attraction] of the before deflection *Flow*-produced acceleration. The magnitude of the reaction acceleration is the same as the magnitude of the deflection.

The ultimate result of the deflection action is the combination of reducing the gravitational attractive acceleration acting on the deflector [and whatever is attached to it] toward the gravitation source plus the introducing of a reactive repulsive acceleration of the deflector [and whatever is attached to it] in the direction away from the gravitation source.

Picturing the Earth with its gravitational field we let "vertical" and "upward" refer to the direction directly away from the gravitating "source". The various individual rays [so to speak] of gravitation, scattered by deflections, all are a combination of a horizontal component and a vertical component, each in various amounts for various rays. The horizontal components cancel out to null. The vertical components total effect after being deflected differs from the pre-deflection rays' vertical total effect and that difference is the overall amount, or magnitude, of deflection.

If every ray's vertical component were curved exactly 90°, i.e. from vertical to horizontal, the total effect of the vertical components after deflection would be zero. Then the overall amount of deflection would be 100% of the natural un-deflected gravitation and the reaction to the deflection would be acceleration equal in magnitude to the natural un-deflected gravitational acceleration but "repulsion" directed away from the "source" rather than "attraction" toward the "source".

The deflection process occurs throughout the length of the deflector crystal. Some rays of gravitational *Flow* are deflected by the first row of atoms of the deflector. Others are deflected by the second row, others the third, and so on. The total deflection is essentially spread linearly uniformly over all of the length of the deflecting crystal.

For the example of every ray's vertical component curved exactly *90°*, i.e. to the horizontal, that would happen linearly uniformly along the crystal length. The result would be that the natural gravitational attraction on the deflector itself would be reduced to 50% of normal.

At the same time the reaction repulsive acceleration magnitude would be 100% of the natural gravitational attraction acceleration because of the overall 100% deflection.

The combined net effect on the deflector itself is then a net "repulsive" acceleration of the deflector of magnitude 100% of the natural pre-deflection attraction offset by the residual gravitational "attraction" on the deflector of magnitude 50% of the natural pre-deflection attraction for it, a net result of repulsion at 50%. The net repulsive acceleration experienced by the deflector would also be experienced by any mass attached to the deflector such as a spacecraft or a flying vehicle.

THE MECHANISM OF THE ANTI-GRAVITATIONAL ACCELERATION

One cannot simply rely on the principle that everything in nature is balanced to account for so dramatic an effect as the repulsive acceleration reaction to the deflection of gravitation – an actual anti-gravity. However, the mechanism producing the effect is simple and natural.

First, as presented in Section 5, natural gravitational acceleration is caused by the *Propagated Outward Flow* from a "source" particle acting on an "encountered" particle. That arriving *Flow*'s μ_0 and ε_0 (scalar quantities not vector and much inverse square reduced relative to their magnitude as originally propagated) combine with the μ_0 and ε_0 in the new outgoing propagation of the "encountered" center (full magnitude as in the *Flow* as originated at the "source"), the combined μ_0 sum and the combined ε_0 sum each being larger values than in the "encountered" particle's originating *Flow*. The result is that that "encountered" particle's new outward *Flow* toward the "source" is slowed relative to its natural otherwise speed per equation 5-2.

Described another way, the *Propagated Outward Flow* from the "source" particle arriving at the "encountered" particle creates a region of an additional increment of μ_0 and ε_0 directly adjacent to the "encountered" particle located on its side that faces the "source".

That increment adds to the natural amount of μ_0 and ε_0 in the natural *Propagated Outward Flow* there from the "encountered" particle causing the "encountered" particle's *Propagated Outward Flow* to be slowed per equation 5-2.

As illustrated in Figures 5-2 and 5-3 and their associated text that effect creates an imbalance in the "encountered" particle's outward momentum propagation and in the consequent reaction back on the "encountered" particle. The "encountered" particle experiences greater reaction to its own outward propagation on its side away from the "source" than on its side facing the "source" the result being acceleration toward the "source".

Second, in the action of the deflector, the components of the incoming vertical gravitational field *Propagated Outward Flow* that are curved away from the vertical by the deflector's atom's are by virtue of that deflection directed over the side of the atom opposite that facing the "source" of the gravitation as depicted schematically in Figure 9-1, below.

Atom

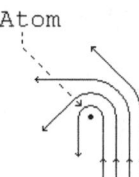

Figure 9-1

That increases the *Propagated Outward Flow* medium concentration on that side of the atom; that is it creates a region of an additional increment of μ_0 and ε_0 directly adjacent to the "encountered" particle but now located on the side of the "encountered" particle away from the "source" rather than the side facing the "source" as in the case of the original un-deflected gravitational *Flow*.

Just as with natural gravitation, that has the effect of reducing the encountered core's propagation in that vertically upward direction that of the increased *Propagated Outward Flow* concentration. That is, the presence *on the opposite side* of a particle of increased medium *Flow* concentration produces the same effect as does natural gravitation's concentration *on the facing side* of a center-of-oscillation, the effect being the same whether the *Flow* is incoming natural gravitation or deflected gravitational *Flow* passing over.

Natural, "attractive" gravitation produces acceleration in the direction of the side of the "encountered" particle where arriving Propagated Outward Flow medium increased the ambient medium μ_0 and ε_0 there slowing natural propagation there. Therefore, the magnitude of the anti-gravitational acceleration is the same as the reduction in the natural gravitational action.

Anti-gravitational "repulsive" acceleration occurs when a same arriving Propagated Outward Flow medium concentration takes place on the side of the "encountered" particle opposite that facing the "source". That is the side of the "encountered" particle to which the deflected incoming Propagated Outward Flow is deflected.

THE MAGNITUDE OF THE ANTI-GRAVITATIONAL ACCELERATION

Every "ray" of *Propagated Outward Flow* encountering the "encountered" particle is either deflected to some extent or not deflected. The magnitude of the gravitational attraction is directly related to the magnitude of the region of an additional increment of μ_0 and ε_0 created directly adjacent to the "encountered" particle and located on its side that faces the "source".

The magnitude of the relative reduction of the natural gravitational attraction that the deflector produces is the relative reduction of the magnitude of the region of an additional increment of μ_0 and ε_0 located facing the "source". Every ray removed by deflection from contributing to that region automatically contributes to the anti-gravitational region on the "encountered" particle's side away from the "source".

Thus, the magnitude of the anti-gravitational "repulsive" acceleration equals the magnitude of the reduction in the natural gravitational "attractive" acceleration.

66

All of the deflected portions of the incoming Flow are deflected toward the region adjacent to the "encountered" particle on its side away from the "source". Therefore, the magnitude of the anti-gravitational acceleration is the same as the reduction in the natural gravitational action. Therefore, the magnitude of the anti-gravitational acceleration is the same as the reduction in the natural gravitational action.

CONTROLLED GRAVITATION SPACE TRAVEL APPLICATIONS

The propulsion system for space travel must perform a number of varied tasks:

- Launch into space from a gravitating source, a planet,
- Soft landing on a gravitating objective, a planet,
- Continuous acceleration for approximately half the total trip,
- Continuous de-celeration for the remainder of the trip,

and provide a reasonable level of planet simulating gravitational environment in the ship.

The propulsion system for a planetary exploration craft must:

- Neutralize local gravitation so that the craft can fly.
- Provide controlled gaining and losing of altitude.
- Provide forward and rearward acceleration, constant velocity cruising, hovering and braking.

and provide a reasonable level of local gravitation within the craft.

In addition both the spacecraft and the planetary exploration craft require substantial electric power to maintain human supportive environment and to power controls, instruments and tools.

These issues are addressed in the following:

Section 10, Gravitation Deflector Engineering Design

Section 11, Applications to the Problems of Space Travel

68

SECTION 10

Gravitation Deflector Engineering Design

| From theory to practical engineering design |

THE DEFLECTOR

The overall deflector consists of:

- A support having a verified perfectly horizontal upper surface for the cubic crystal deflector bottom face to rest upon;

- Monolithic Silicon cubic crystal ingots as follows:

    · *30 cm* in diameter,
    · *50 cm* or more thick,
    · with the orientation of the cubic structure marked for proper placement of tilt-generating shims, and
    · with the bottom face of the cylinder sawed and polished flat at a single cubic structure plane of atoms.

    [or equivalent tilt orienting / calibrating provisions]

- Precision shims *4.5 mm* thick for producing the tilt of the cubic crystal ingot, the shims located at the mid-point of two adjacent sides of the horizontal plane of the cubic structure as in Figure 10-1 below.

- Alternatively, a precision tilt-generating mechanism.

- For an array of ingots for a larger area than a single ingot can provide, the individual ingots can be machined to fit snugly together. That could be done by machining them to a square cross section or, better, to a hexagonal one.

69

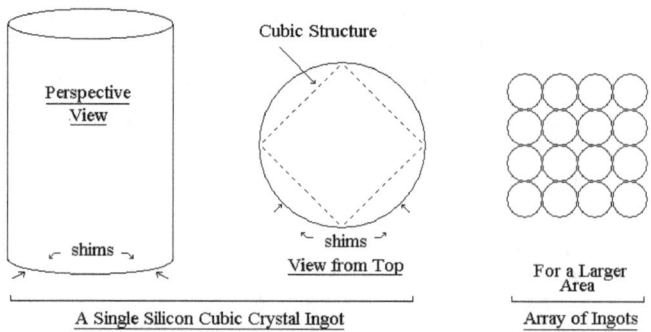

Figure 10-1 – The Silicon Cubic Crystals Arrangements

PRACTICAL ASPECTS IN DESIGN ENGINEERING

While the net Earth gravitational field is vertically upward, i.e. radially outward from the Earth's surface, local gravitation is radially outward from each particle of matter. As in Figure 10-2 below, a mass above the Earth's surface receives rays of gravitational attraction from all over the Earth's surrounding surface and from the underlying body of the Earth.

The net effect of all of the rays' horizontal components is their cancellation to zero. The net effect of all of the rays' vertical components is Earth-radially-outward vertically acting gravitation.

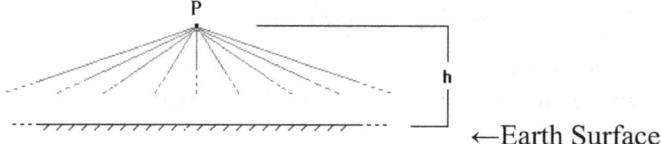

Figure 10-2 - Rays of Gravitation from the Surroundings

Gravitational Ray's Horizontal and Vertical Components.

One can consider all of the net gravitational effect on objects as being due to the vertical component of all of the myriad rays of gravitational field *Flow* at a wide variety of angles to the horizontal.

The various rays of the *Flow* propagation from the individual particles of the gravitating body [for example the Earth] are from each individual particle of it to the selected point [above the gravitating body] on which their action is being evaluated. That is the point P in the above Figure 10-2.

The Earth's gravitational action along a ray of *Flow* takes place from the Earth's surface to deep within the Earth. The inverse square effect, that the strength of a *Flow* source is reduced as the square of the increase in the radial distance of it from the object acted upon, is exactly offset by that the number of such sources acting [per "ray" so to speak] increases as the square [non-inverse] of that same radial distance. That is, the volume, hence the number, of *Flow* sources for a ray of propagation at the object is contained in a conical volume, symmetrically around the ray with its apex at the object acted upon.

However, because the net gravitational effect is produced only by the vertical component of each ray of *Flow* propagation, the effectiveness of each ray is proportional to the Cosine of the angle between that ray and the perfectly vertical as the angle θ in Figure 10-3 below.

Figure 10-3 – The Gravitational Field Ray Angle to the Vertical

The actual total gravitational action includes all rays from $\theta = 0$ through to $\theta = 90^\circ$. That would require an infinitely large deflector to act on all such rays, a disk of infinite radius. For lesser values of the maximum θ addressed, the portion of the total gravitation sources included is the integral of $Cos[\theta] \cdot d\theta$ from $\theta = 0$ to $\theta = Lesser\ Value$. The integral of the *cosine* is the *sine*. Example lesser portions of the total gravitational action addressed as θ varies are presented in Table 10-4 below.

θ	$Sin[\theta]$ = Fraction of Total Maximum Gravitational Action
0°	0.000
30°	0.500
45°	0.707
60°	0.866

Table 10-4

The gravitational deflector as a disk beneath the *Object* to be levitated must extend horizontally far enough to intercept and deflect the $Chosen\ Lesser\ Value$ of angle θ rays of gravitational wave *Flow* that are able to act on the *Object* of the deflection as depicted in Figure 10-5 below.

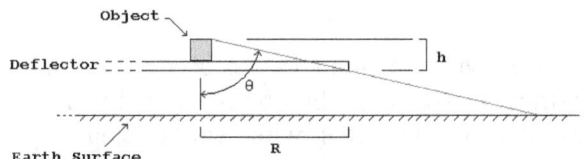

Figure 10-5 – Size Requirements for a Disk Shaped Deflector

For the perfectly vertically traveling rays of gravitation waves the required vertical distance that must be traveled within the cubic crystal is the previously presented at least $50\ cm$ and 0 horizontal distance is traversed in so doing. But a ray at angle θ, in order to traverse the required 50 cm vertically, must traverse horizontally $50 \cdot Tan[\theta]\ cm$, at the same time. For θ more than 45° that can become quite large and the deflector likewise.

Because the deflector disk must extend over a large area to deflect most of the gravitation, an alternative, and better, solution to the problem of rays of gravitation arriving over the range from $\theta = 0\ to\ \theta = 90^\circ$ is to wrap the deflector up the sides of the *Object* to be levitated as shown below.

71

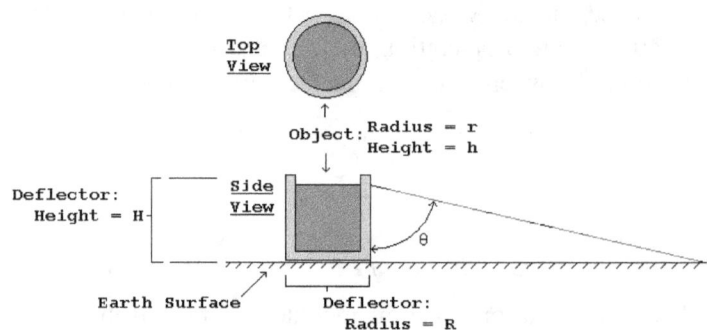

Figure 10-6 – More Efficient Cup-Shaped Deflector

In this configuration the deflector takes up little more space than the *Object* levitated. However, the non-perfectly vertical traveling rays must still travel within the cubic crystal the horizontal distance *50·Tan[θ] cm*. That requires that the horizontal thickness of the vertical sides of the cup-shaped deflector must be of that *50·Tan[θ] cm* thickness.

Because the value of *Sin[θ]* and, therefore, the fraction of the total gravitational action, increases relatively little above *θ = 60º* whereas the value of *Tan[θ]* increases quite rapidly, from *1.7 to ∞* above *θ = 60º* that *θ = 60º* is the appropriate value to which to design. The thickness of the "walls" of the "cup" would then be *50·Tan[60º] = 85 cm*. The deflector would be only slightly larger than the *Object* levitated.

Gravitation Deflector Design Parameters

The Deflector is a cup shaped array of monolithic Silicon cubic crystals. The crystals forming the flat "base" of the "cup" must be *0.5 m* in height. The "sides" of the "cup" will be the same kind of *0.5 m* crystals stacked and aligned vertically. The thickness of the "sides" must be *0.85 m*.

The crystals are grown with circular cross-section and in diameters up to *30 cm*; however, those cylindrical pieces must then be machined to hexagonal or square cross section for a number of them to fit together with negligible open space. The cross-section area of these crystals is $\pi \cdot d^2 / 4 = 0.785 \cdot d^2$

For a circular deflector the configuration is poorly compatible with arranging the crystals in a close-fitting array unless it involves a large number of crystals each of small cross-section relative to the horizontal cross-section of the overall deflector. For that case the crystals should be machined to hexagonal cross-section. For smaller deflectors the configuration should be rectangular and the crystals machined to square cross-section.

Case	Preferred Crystal Cross-Section	Crystal Cross-Section Area	Percent Used of Original Crystal
Circular Deflector	Hexagonal	$[\sqrt{3}/3] \cdot d^2 = 0.577 \cdot d^2$	73.5
Rectangular Deflector	Square	$d^2/2 = 0.500 \cdot d^2$	63.7

Table 10-7

a. Circular Cross-section Gravitation Deflector Structure

A circular cross-section gravitation deflector structure to provide deflection for an object of height, h, and diameter, d meters would have the following parameters.

```
Base Disk: Thickness = t = 1 Crystal Layer = 0.5 m
           Diameter = d + 2·[t = cup sides thickness]
           Area = π·[d + 2·t]²/4 = 0.785·[d + 1.7]²
```

Cup Sides:

```
    Thickness              t = 0.85 m
    Outside diameter [OD] = d + 2·t = d + 1.7
    Inside diameter  [ID] = d
    Height               = h
    Height Nr. of Layers = h/0.5
    Area of Layer        = π·[OD² – ID²]/4
                         = 0.785·[OD² – ID²]
```

Taking Silicon at $1.00\ \$/kg$ and its density at $2,329\ kg/m3$ the examples below obtain [MKS units and $1\ m = 39.37"$]. The $0.85\ m$ thickness of the "cup" "sides" requires 20 layers horizontally of $2"$ crystals.

d	h	Cup Disk Base		Cup Sides		Total Volume	Total Cost $	Nr. of 2" Hex Crystals
		Area	Volume	Area	Volume			
1	1	5.72	5.75	4.94	4.94	10.7	24,897	13,280
10	10	131	1310	29	290	319	742,951	779,570

Table 10-8

b. Square Cross-section Gravitation Deflection Structure

A square cross-section gravitation deflector structure to provide deflection for an object of square cross-section side, s, and height, h meters would have the following parameters.

```
Base Square: Thickness = t = 1 Crystal Layer = 0.5 m
             Side = s + 2·[t = cup sides thickness]
             Area = [s + 2·t]² = [s + 1.7]²
```

Cup Sides:

```
    Thickness              t = 0.85 m
    Outside square side  OS = s + 2·t = s + 1.7
    Inside square side   IS = s
    Height               = h
    Height number of Layers = Height/0.5
    Area of Layer        = OS² – IS²
```

Taking Silicon at $1.00\ \$/kg$ and its density at $2,329\ kg/m3$ the examples below obtain [MKS units and $1\ m = 39.37"$]. The $0.85\ m$ thickness of the "cup" "sides" requires 3 layers horizontally of $12"$ crystals.

s	h	Cup Disk Base		Cup Sides		Total Volume	Total Cost $	Nr. of 12" Square Crystals
		Area	Volume	Area	Volume			
1	1	7.3	7.3	6.3	6.3	13.6	31,674	195
10	10	137	1370	36.9	369	1,739	4,050,131	2,486

Table 10-9

Calibrating the Individual Silicon Crystals

The individual crystals making up the deflector cannot be grown exactly identical to each other. In each the orientation of the long axis of the cubic crystal structure may vary minutely from each of the others. That is, it is not certain that each crystal's base is purely a single plane of atoms of the cubic structure and thus is exactly perpendicular to the long axis of the crystal.

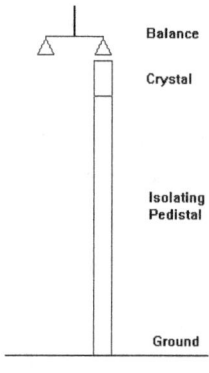

To find the optimum tilt and orientation for a single crystal the tilt must be varied over the range of possibilities while the effect of gravitation from exactly below it is observed on a balance scale. But, most of the effect of gravitation on a single crystal is not from exactly below.

The solution to that problem is to conduct the optimization atop a structure that, relying on the inverse square effect, effectively isolates the crystal from most of the gravitation from surrounding sources except that exactly below it – a high pedestal having a cross section comparable to that of the crystal, Figure 10-10.

Figure 10-10

To conduct that calibration on thousands of crystals should not be necessary if a method can be developed to exactly measure the long axis orientation in any given crystal. The process can then determine the optimum orientation of the crystal tilt relative to the actual long axis of a few cubic crystals being calibrated. That same crystal tilt relative to the actual long axis can then be applied to each of the other crystals.

The long axis orientation problem could also be solved by a method of insuring that the base of each crystal is a single plane of atoms of the cubic structure.

Alternative to Calibration

Monolithic silicon cubic crystals are commercially available with the ends nearly a single plane, that is within *0.2 degrees* of the *(100)* plane of the cubic structure. In view of the various effects analyzed in Appendix D, and their resolution in its section *The Random Distribution Solution to The Crystal Tilt*, that amount or moderately more of inaccuracy in the crystal tilt is of no significance except that it potentially may call for crystal thicknesses moderately greater than *0.5 m*.

Before the design can further progress to definitive deflector structures and the control mechanisms for them two other actions that are beyond the scope of this work are required:

1. Experiments and testing to accurately establish the various overall design parameters: *e.g.* minimum required crystal dimensions and effectiveness of various tilts.

2. Specific design decisions are required for each of the various applications of gravitation control described in the next following Section 11: e.g. spaceship, planetary surface flying vehicle and power plant.

SECTION 11

Conceptual Gravitational Applications

A GRAVITATION DEFLECTOR SPACECRAFT DEEP SPACE DRIVE

A spacecraft gravitation deflector drive would be a deflector in cup form, mounted on the "rear" of the spacecraft and extending the spacecraft's full length to its "nose". In the case of the invention of the automobile the motor was instinctively placed in front of the cab where the horses it replaced were located the cab being otherwise unchanged. That was done even though the motor's function was to drive the rear wheels because driving the front wheels with steering was too complex.

Likewise, the initial or instinctive concept of a gravitation deflection spacecraft places the deflector at the rear or bottom of the space capsule where the rocket engines it replaces were located The concept is of a rocket with its narrow aerodynamic rocket form but with different engines as in Figure 11-1, to the right.

Legend: Deflector   Spacecraft   Figure 11-1

However, the nature of gravitation is such that the spacecraft can be as large as we may wish and there is no need for aerodynamic considerations for flying in atmosphere. The spacecraft will be a human habitation for a long time, even for years, and must be compatible with human needs. Thus the spacecraft should resemble the kind of environment that humans experience in every day life: office buildings and community residences. Those factors would appear to call for a spacecraft configuration more as that in Figure 11-2, below. Such a craft could be multi-level [multi-storey] with levels or "decks" for: residences, administration, laboratories, work shops and a garage for planet surface flying vehicles. The top level could offer a beautiful "sky view".

Legend: Deflector   Spacecraft

Figure 11-2 – A Gravitation Deflector Driven Spacecraft

77

This configuration would satisfy a number of functions. The deflector would provide [all without use of fuel]:

- Launching of the spacecraft vertically upward at an upward acceleration of approximately one-half of local natural gravitation, for Earth an upward acceleration of about 16.1 $ft/s2$;

- Landing and re-launching of the spacecraft at any gravitating body such as the Moon, Mars or Proxima Centauri b;

- Deep space transit propulsion between gravitating bodies;

- Partial protection from deep space radiation and cosmic ray particles by virtue of the ½ to almost 1 meter thickness of the Silicon deflector;

- A gravity environment within the spacecraft of zero natural gravitation plus an artificial gravitation due to the acceleration of the ship in whatever amount that it is at any particular time [taking "down" as toward the deflector end of the ship].

The engineered arrangements for varying the amount of deflection so as to vary the acceleration would be means of controlled changing of the orientation of selected portions of the Silicon cubic crystals. The engineered arrangements for varying the direction or orientation of the spacecraft would be a 3-axis system of angular momentum wheels

For a spaceship in free space the gravitational *Flow* environment is different from on Earth. In the case of only one gravitation source near enough to be of any important effect departing such a source after launch from it requires simply aiming the stern of the ship toward that source. Controlled landing on it requires simply aiming the stern of the ship toward that source and controlling the acceleration by varying the deflection.

In general, however, in deep inter-planetary space gravitation is present albeit fairly weekly because of inverse square reduction of intensity, and it is present in various amounts with attraction toward various differently located sources. As with the sailing navigation using the wind in earlier centuries, spaceship travel within the Solar System may require techniques analogous to: sail craft's tacking on various headings, "crabbing" into partial "cross wind" as aircraft do, and in general going "where the winds permit". In the spacecraft case the "winds" are the various direction gravitational *Flows* available from which to generate acceleration and to which the spacecraft is subject to attraction.

Solar System navigation is further complicated by the destination's continuous motion. The navigation must be toward where the destination will be upon spacecraft arrival at it as compared to where the destination currently is.

For inter-stellar navigation there is the possibility of near light speed travel. The deflector could provide continuous, fuel-less acceleration to the spacecraft throughout its trip. The continuous acceleration would accelerate the craft during the first part and, with the craft re-oriented using the 3-axis system of angular momentum wheels, decelerate the craft for approach to the destination.

Because the acceleration is independent of the mass of the spacecraft its mass could be quite large and able to carry everything needed for an extended trip and for survival at the destination. The spacecraft could therefore also have whatever large amount of shielding is needed against the radiation hazards of free space.

A DEEP SPACE TRIP FROM EARTH TO THE MOON

For launch from Earth the spacecraft starts from resting on its stern, a position to which it most recently arrived. The launch is propelled by anti-gravitational repulsing of the spacecraft relative to the Earth.

The step between the launch and the aimed direction toward the Moon is an orbit of the Earth. Once initially in that orbit the orbit can then be shifted to the correct orientation for the spacecraft to leave the orbit headed toward the Moon.

The spacecraft remains in Earth-repulsing mode [its stern toward the Earth] for approximately half of the trip to the Moon, that is to the point where the Moon's attraction for the spacecraft takes over from the Earth's attraction for the spacecraft.

At that point the spacecraft changes its orientation so that its stern is toward the Moon. The spacecraft is then in the mode of repulsing the Moon toward which the spacecraft has a velocity accrued from its Earth-repulsing stage of the trip.

The magnitude of the Moon repulsing action is varied so as to guide the spacecraft to a soft landing on the moon.

A capability is required for selecting a location in which to land on a space travel destination and for translating the spacecraft so as to land at that destination. Those problems are treated in the following.

A GRAVITATION DEFLECTOR PLANET SURFACE FLYING VEHICLE

A gravitation deflector flying vehicle would have a deflector in cup form, underneath the payload compartment of the vehicle. That "elevation deflector" is for the purpose of providing vertical acceleration and maintaining travel levitation as in Figure 11-3 below.

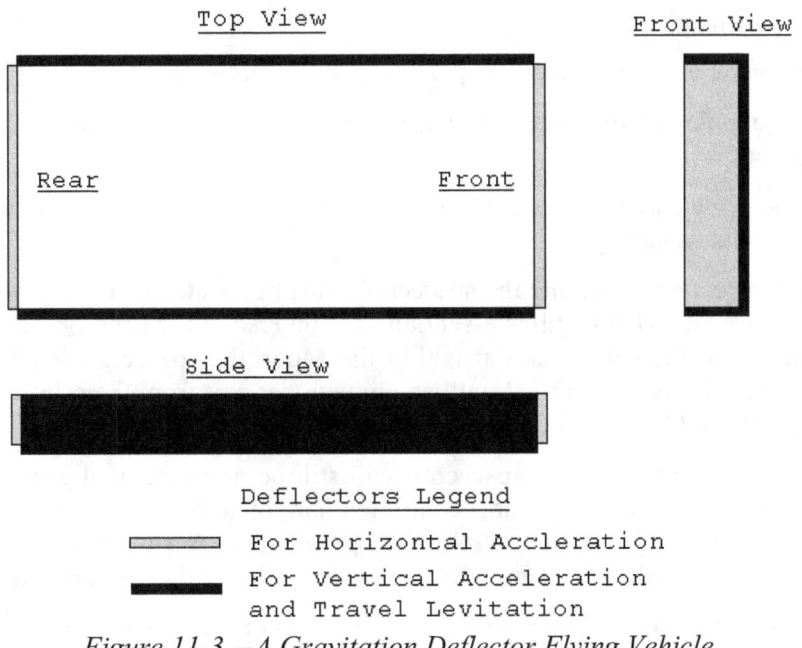

Figure 11-3 –A Gravitation Deflector Flying Vehicle

In addition to that elevation deflector the vehicle would have a front and a rear deflector for the purpose of providing horizontal acceleration and braking by acting on the horizontal components of the ambient gravitational field. The flying vehicle differs from the form for a spacecraft in:

- needing only modest acceleration capability vertically upward beyond sufficient to maintain its constant altitude levitation,

- employing means to generate horizontal acceleration while maintaining vertical levitation.

The vehicle requires a 3-axis system of angular momentum wheels for controlling its horizontal direction and maintaining its positional attitude in space.

This deflector configuration [all without use of fuel] will:

- Provide controlled vehicle levitation for take-off, landing, and travel,

- Provide controlled horizontal propulsive acceleration and "braking",

- But there is the problem of sufficient gravity for the passengers.

The vertical acting deflectors cannot provide artificial gravity by the action of vertical acceleration because the vertical acceleration is controlled to only maintain levitation at a given altitude except for take-off and landing. However, maintaining levitation requires significantly less than 100% vertical deflection. If, for example, levitation required only 50% vertical deflection then the gravitation within the vehicle would be the remaining undeflected 50% of natural gravitation.

A MOON LANDING AFTER TRAVEL TO THE MOON

Upon arrival at the Moon the first action is to reconnoiter and select a suitable landing cite, one that:

- Is convenient for the planned purpose and actions after landing, and

- Is suitable for the size and structure of the spacecraft in terms of the site's geography.

For that purpose the spacecraft launches one or more of its Gravitation Deflection Flying Vehicles to conduct a survey.

Upon selection of the site the spacecraft must be located so that it can land there by a vertical descent. That requires essentially the inverse of the orbit process employed at the beginning of the trip. Upon arrival at the Moon the spacecraft is entered into a lunar orbit. The orbit is then shifted until so aligned that exit from the orbit will result in arrival at the selected landing site.

In spite of that action the spacecraft will still be imperfectly aligned and located for the landing. To adjust that requires controlled horizontal translation of the spacecraft. As is the case with large ocean vessels, the spacecraft is too massive, has too much inertia, to maneuver itself into exactly the intended location and configuration.

The solution to that is analogous to the seaport operations for large vessels – tug boats maneuver the large vessels into position. In the case of landing on the Moon or any planetary body the tug boats are the same Gravitation Deflector Planet Surface Flying Vehicles that performed the earlier reconnaissance.

GRAVITO – ELECTRIC POWER GENERATION

Electric power for the controls, lights, instruments and temperature control in the Gravitation Deflector Planet Surface Flying Vehicles can be supplied by re-chargeable batteries, the batteries being on charge whenever the vehicle is parked in the vehicle garage on the main spacecraft. But the electric power for that charging and for all of the electric needs of the main spacecraft must be generated on board that spacecraft. The means for that is gravito-electric power generation.

Gravito-electric power generation is similar to hydro-electric power generation in which the energy of water falling in Earth's gravitational field powers water-turbines that drive electric generators.

In gravito-electric power, depicted schematically in the figure below, a gravitation deflector repulses the water in the central region of the mechanism. That water is effectively lighter than that in the outer region, which is acted on by whatever is the local gravitation for the benefit of the humans on the craft. The lighter water floats up on the in-flow under it of the heavier natural gravitation water. The result is continuous circulation of the water, like a continuous waterfall.

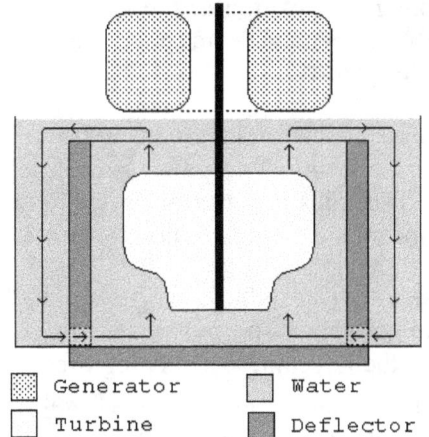

| Generator | Water |
| Turbine | Deflector |

Figure 6-1 – A Gravito-Electric Generator

Water turbines like those used in hydro-electric plants can be placed in the gravito-electric continuous water flow to drive electric generators as in hydro-electric plants.

Gravito-Electric power plants could replace all Earth-based fossil fuel power plants for a major contribution toward relieving the problems of global warming.

SPACE NAVIGATION

[The content of this discussion of space navigation is based on a large variety of materials that will be found on Wikipedia by searching on "space navigation".]

While it might seem that navigation within the Solar System and between stars should be quite simple because one can "see" both the source and the destination the problem is actually much more complicated. Within the Solar System the motions of the various destinations are very significant and travel among them requires going to where the destination will be at the time of the travel arrival there. While that factor is less significant for interstellar travel, interstellar travel involves accelerating for the first "half" of the trip and decelerating for the remainder so that it is essential to know exactly where one is located relative to the source and the destination so that the acceleration phase and the deceleration phase can both be properly managed.

With men having been placed in space in the International Space Station and the advent of on-going planning for future human travel to Mars, interest in the problem of space navigation has developed. In 2003 the European Space Agency studied the feasibility of pulsar navigation and in 2012 two different pulsar navigation studies were conducted, one by GMV Aerospace and Defence (Spain) and one by the National Physical Laboratory (United Kingdom).

In 2016 the People's Republic of China launched an experimental pulsar navigation satellite for the purpose of characterizing 26 nearby pulsars for their pulse frequency and intensity to create a database for navigation for future operational missions.

Then, in 2018 NASA conducted a pulsar navigation experiment called SEXTANT (Station Explorer for X-ray Timing and Navigation Technology) using the International Space Station. That pulsar navigation system demonstrated an accuracy of within 7 kilometers in defining the position of the space station. While in terms of the space station orbit that accuracy may seem poor, accuracy within 7 kilometers is essentially spot on for interstellar travel.

A pulsar is a rotating star that emits a narrow beam of radiation. That radiation can be observed only when the beam of emission is pointing toward Earth. The rapid rotation of the star, from milliseconds to seconds per cycle of rotation, is responsible for the corresponding pulsed appearance of its emission.

The periods of pulsars make them very useful tools for astronomers. Each pulsar's rate of rotation and corresponding pulse rate are very stable. Certain types of pulsars rival atomic clocks in their accuracy in keeping time.

Pulsar navigation is a technique where the periodic X-ray signals emitted from pulsars are used to determine the location of a spacecraft, such as a spacecraft in deep space. The spacecraft would compare received pulsar X-ray signals with a database of known pulsar frequencies and locations. Similar to GPS, that comparison allows the spacecraft to triangulate its position accurately.

Just as sailors look to the stars to navigate and determine their position, interstellar travelers may use pulsars to navigate the universe.

And

Just as the sail-driven ships of past centuries explored the world with fuel-free travel by controlled use of the wind;
The new gravitation technology will enable fuel-free exploration of space by control of the ubiquitous gravitational field.

Appendix A

Why No Immediate Mutual Annihilation

BACKGROUND OF THE PROBLEM

The Big Bang could only have resulted in equal amounts of matter and antimatter for the sake of the principle of conservation as presented in Section 2, *The Origin of Matter Is the Origin of Gravitation* with the assumption that there would have been a complete and almost instantaneous mutual annihilation.

Because that annihilation did not take place it has been hypothesized that the original symmetry was slightly skewed in favor of matter and that the universe is now all matter, all original antimatter having been annihilated with an equal amount of original matter. However that skewed balance conflicts with conservation in the Big Bang.

The Big Bang had to produce equal amounts of matter and antimatter and their total mutual annihilation did not occur because of the conditions there. Rather, while a moderate amount of initial matter / antimatter mutual annihilations may have taken place our present universe contains the remaining matter and antimatter in equal amounts, between some particles of which further mutual annihilations still occur at a modest rate.

The failure of comprehensive matter-antimatter immediate annihilation to occur develops as follows.

CONDITIONS AFFECTING MATTER / ANTIMATTER MUTUAL ANNIHILATION

What Is a Matter / Antimatter Annihilation ?

A positron-electron mutual annihilation, for example, is

(A-1) $_1e^0 + {}_{-1}e^0 \Rightarrow \approx + \approx$ where $\approx$ is a photon of gamma radiation

It happens as follows [per equation 2-6].

(A-2) $({}_1e^0) + ({}_{-1}e^0) = U_c \left[1 - Cos(2\pi \cdot f_e \cdot t)\right] - U_c \left[1 - Cos(2\pi \cdot f_e \cdot t)\right]$

$= 0$

The two oscillations literally cancel. The annihilation occurs because the two are point-by-point inverses of each other. Such an annihilation is depicted in Figure A-1 on the following page.

In general for a particular particle and some particular anti-particle of it, their phases and frequencies will not be identical because of their different velocities and histories of relativistic frequency shifts. However, for them to mutually annihilate they must remain co-located for some brief moment sufficient for the event to occur.

For the particles to be co-located for a brief moment their positions and velocities must be identical, which means that their frequencies and their phases will also be identical.

The mutual annihilation energy is the conversion into energy of the entire mass of the two particles involved. The mass of each of the particles is its oscillation [there is nothing else to be the mass]. At annihilation the two particles' oscillations cease to exist by cancelling each other out. Since the center oscillations cease, the last waves of *Propagated Outward Flow* are followed by no flow at all from those centers.

E-M radiation is the propagation of changes in the *Propagated Outward Flow*, changes usually caused by velocity changes of charged particles. The ceasing at annihilation of the oscillations of the two particles involved [the largest change possible] causes a pair of gamma photons, equation $A-1$, to be propagated.

The photons carry off conservation maintaining energy and momentum. The frequency of each photon is the frequency of the oscillation that just ceased, which corresponds to the mass of the particle. In other words the photon energy, $W = h \cdot f$, is the energy equivalent of the entire mass of the of the particle annihilated.

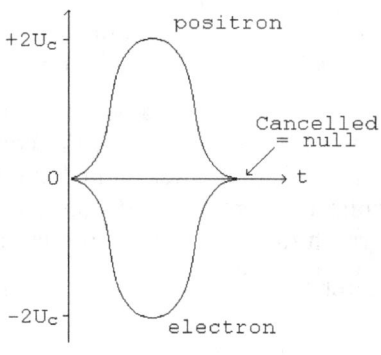

Figure A-1
A Mutual Annihilation

The first issue to investigate is the necessary conditions for a matter / antimatter annihilation to take place: how close must the particle and its antiparticle be and for how long must they remain in such sufficiently intimate contact ?

In addition to those two factors there is the more obvious requirement that the two particles involved be true antiparticles of each other [for example, a proton and an antiproton or an electron and a positron, but not a proton and a positron nor a proton and an electron]. Furthermore in general, particle / antiparticle annihilations are relatively unlikely between electrically neutral particles [for example, a neutron and an antineutron] because the only effects tending to bring the two together are their very weak gravitational attraction or chance encounter.

The Closeness Criterion

Indication of how close the two participating particles must be for their annihilation to take place can be found from the decay of a free neutron [not one that is part of an atomic nucleus] into a proton and an electron, a natural process with a mean lifetime before decay of about 881.5 seconds. For the neutron decay to be successful the proton and electron product particles must derive from the parent neutron not only their rest masses but also sufficient kinetic energy so that they are at escape velocity relative to each other, else they would be attracted back together and recombine. [One can neglect the also emitted electron anti-neutrino which is of zero or negligible mass.]

The escape velocity of the two particles is, at first consideration, an awkward problem because the separation distance of the two particles, which appears in the denominator of the expression for their Coulomb attraction, would seem to be required to be as small as zero. That is, at first consideration the escape velocity required is infinite. But, since infinite escape velocity is impossible yet the escape occurs, then the starting point, the minimum separation distance that can occur must be greater than zero. In other words, the neutron decay products, a proton and an electron, exist as such only when separated by some minimum Separation Distance, *s*, and their state at lesser separation distances appears as their parent neutron.

Therefore, since if the proton and the electron are separated by less than that minimum distance they do not exist as proton and electron but rather as the neutron, and at separation distances greater than that minimum they are the pair of separate particles, then that Separation Distance is a measure of how close a proton and an electron must be to unite into a neutron and is indicative of the spacing at which a particle and its antiparticle mutually annihilate.

The point is that the excess of the mass of the neutron over that of a proton plus that of an electron must supply the proton and electron relativistic kinetic masses needed to escape the decaying neutron. The detailed analysis and relativistic calculations can be found in Appendix A-1, *The Neutron*. The results are as follows.

(A-3) - The escape velocities:

$$v_e = 275,370,263. \text{ meters per second}$$

$$= 0.918,536,33 \cdot c$$

$$v_p = 379,350.6975 \text{ meters per second}$$

$$= 0.001,265,378 \cdot c$$

- The minimum Separation Distance:

$$S = 1.3 \cdot 10^{-15} \text{ meters}$$

Some years ago experiments involving measurement of the scattering of charged particles by atomic nuclei, yielded an empirical formula for the approximate value of the radius of an atomic nucleus to be

(A-4) Radius = $[1.2 \cdot 10^{-15}] \cdot$ [Atomic Mass Number] meters

which formula would indicate that the radius of the proton as a Hydrogen nucleus (atomic mass number $A = 1$) is about $1.2 \cdot 10^{-15}$ meters.

The mass of the proton can be expressed as an equivalent energy, $W_p = m_p \cdot c^2$, and that as an equivalent frequency, $f_p = m_p \cdot c^2 / h$, or as an equivalent wavelength, $\lambda_p = c / f = h / m_p \cdot c$. That wavelength (not a "matter wavelength") for the proton is

(A-5) $\lambda_p = 1.321,410,0 \cdot 10^{-15}$ meters

quite near to the empirical value for the proton radius from equation *(A-4)* and the Separation Distance, *S*, of equation *(A-3)*. Thus the Separation Distance boundary between a proton and an electron as separate particles versus combined into a neutron is about *1* proton radius, the equivalent wavelength for the proton mass per equation *(A-3)*.

Then for a proton and an antiproton the boundary between their being the two separate particles and their mutually annihilating is a proton radius, a Separation Distance of $S_p = \lambda_p = 1.321,410,0 \cdot 10^{-15}$ *meters*. At that boundary if their velocities have a sufficient net component directly toward each other [per the time criterion, below] they would seem to be able, and likely, to mutually annihilate, and otherwise the annihilation would seem not possible.

Similarly, the mass of the electron or the positron can be expressed as the equivalent energy, $W_e = m_e \cdot c^2$, and that as its equivalent frequency, $f_e = m_e \cdot c^2 / h$, or equivalent wavelength, $\lambda_e = c/f = h/m_e \cdot c$. That wavelength (not a "matter wavelength") for the electron / positron is

(A-6) $\quad \lambda_e = 2.426,310,6 \cdot 10^{-12}$ *meters*.

Then for an electron and a positron the boundary between their being the two separate particles and their mutually annihilating is a Separation Distance of $S_e = \lambda_e = 2.426,310,6 \cdot 10^{-12}$ *meters*. At that boundary if their velocities have a sufficient net component directly toward each other [per the time criterion, below] they would seem to be able, and likely, to mutually annihilate, and otherwise the annihilation would seem not possible.

Then, what is that sufficient net velocity ?

The Time Criterion

The mutual annihilation of a particle and its antiparticle is symbolized as in the following example for a proton and an antiproton.

(A-7) $\quad _1p^1 + _{-1}p^1 \Rightarrow \gamma + \gamma \quad$ where γ is a gamma photon

In the present case of a proton and an antiproton the mass of each of the protons is converted into the energy of the related γ photon. The frequency and period of each of those two photons is as follows.

(A-8) $\quad f_{\gamma p} = m_p \cdot c^2 / h$

$\qquad T_{\gamma p} = 1/f_{\gamma p} = h/[m_p \cdot c^2] = 4.407,749,3 \cdot 10^{-24}$ *seconds*

In communications theory it is shown that a sinusoidal oscillatory signal must be sampled at least twice per cycle for the signal to be correctly represented. That is, two independent datums are required so as to determine the value of the oscillation's two absolute parameters, its amplitude and its frequency. [It's phase is relative, not absolute.] That implies that the time duration of a proton / antiproton mutual annihilation must be the period of each of the resulting photons.

(A-9) $\quad \Delta t_{proton / antiproton} = T_{\gamma p} = 4.407,749,3 \cdot 10^{-24}$ *seconds*

Similarly for an electron / positron mutual annihilation, the time duration would be

(A-10) $\Delta t_{electron / positron} = T_{\gamma e} = 8.093,301,0 \cdot 10^{-21}$ seconds.

While those are very brief times they are not instantaneous.

In the case of a particle and its antiparticle coming together from significantly far apart, the particles will have accumulated significant velocity toward each other by the time they arrive at Separation Distance s because of having been accelerated by their mutual Coulomb attraction. However, the situation was different for the Big Bang.

WHY THE CRITERIA FAILED IN THE CASE OF THE BIG BANG

The number of particles resulting from the original Big Bang is estimated to have been about 10^{85} [Appendix B, *The Limitation of the Original Envelopes*], and those particles emerged on paths that were initially radially outward. The event was overall spherically symmetrical on the large scale, but at the local particle level perfect symmetry was impossible because of the nature of finite particles versus a smooth non-particulate substance. Initially all of the particles were on divergent paths although for two adjacent particles the amount of the divergence was minute.

For a proton and an adjacent antiproton in the Big Bang to be separate [not annihilated] at the instant of being projected outward in the Big Bang, they had to be separated by at least the above-developed $s_p = 1.321,410,0 \cdot 10^{-15}$ *meters*. For them to then annihilate their Coulomb attraction would have had to accelerate them into co-locating in the required time criterion starting from their initially zero velocity toward each other. [Actually they would have had non-zero but minute velocities away from each other because each follows its own outward radial path.] The issue is whether their Coulomb attraction can accelerate the two particles to the point of co-locating within the time frame of equation *A-9* [or equation *A-10* for an electron / positron case].

If, for example, for their mutual annihilation, the proton or the antiproton is to travel <u>at constant velocity</u> its half of the separation distance, $\frac{1}{2} \cdot s_p$, in time $T_{\gamma p}$, so as to be co-located with its antiparticle at the end of that time, it would require a speed of

(A-11)
$$ v_p = \frac{\frac{1}{2} \cdot s_p}{T_{\gamma p}} = 0.5 \cdot c \quad \text{[half light speed]} $$

and if the electron or the positron, for their mutual annihilation, is to travel its half of the separation distance, s_e, in time $\frac{1}{2} \cdot T_{\gamma e}$ <u>at constant velocity</u> it would require a speed of

(A-12)
$$ v_e = \frac{\frac{1}{2} \cdot s_e}{T_{\gamma e}} = 0.5 \cdot c \quad \text{[half light speed]}. $$

The achieving of that speed, if even only by the very end of the extremely short time period of the acceleration and travel, 10^{-21} *seconds or less*, would be difficult. The particles moving continuously at that <u>constant velocity</u> throughout their travel from separated to co-located is impossible in that they commence their travel of distance s from essentially zero velocity toward each other.

Finally, the posited particle and its antiparticle, emerging from the Big Bang, with spacing adjacent to each other as closely as possible, and on radially outward paths, were not alone. They were surrounded by a more or less uniform, symmetrical, large group of like particles and antiparticles. Any Coulomb tendency to unite the posited particle pair was largely offset by the similar tendency of each member to unite with the adjacent particle on its other side. The net Coulomb action on a specific particle or antiparticle was certainly insufficient to produce enough acceleration to enable the particle to transit its half of the Separation Distance in the required gamma photon period.

In summary:

- Adjacent Big Bang product particles and their antiparticles,

- Initially spaced optimally for co-locating [as closely as possible yet independently separate],

- Traveling outward at near light speed on essentially parallel paths [actually minutely diverging paths],

- Are unable to accelerate toward each other, from zero initial such velocity, quickly enough for their annihilation to produce the known actual gamma photons that would have to result from their mutual annihilation.

- That is, they cannot travel to the point of annihilation in time for the annihilation gamma photons to be the correct frequency to carry off the energy equivalent of the input particles, the pre-annihilation proton / antiproton or electron / positron.

In other words a Big Bang mutual annihilation was much more difficult, and rare, than one might have assumed. A large scale annihilation of matter and antimatter could not have taken place in the Big Bang. The result is that the present universe contains both matter and antimatter in equal amounts because of the original symmetry.

A UNIVERSE CONTAINING BOTH MATTER REGIONS AND ANTIMATTER REGIONS

Why Matter and Antimatter Regions Are Able to Co-Exist

Of course, matter / antimatter mutual annihilations in general are not as awkward as they were for the original Big Bang with its peculiar initial conditions. Of interest here, however, is the case of the interstellar medium. It is the interstellar medium that must be examined because it is the natural boundary between regions of matter and regions of antimatter; where, if they are to occur, the anticipated matter / antimatter annihilations should be occurring and yielding their looked-for gamma ray flux.

In the interstellar [and intergalactic] medium the particles and antiparticles start from being significantly separated, residing in the vacuum of interstellar space, which vacuum, while not devoid of competing particles, has a much lower particle density than the original Big Bang. They do not suffer the disadvantage of being in a dense milieu of particles and antiparticles whose Coulomb attractions tend to cancel out their effects. And, they avoid the disadvantage of always starting their mutual Coulomb attraction toward each other with no initial velocity. Without regard for any mutual attraction between particular particles and antiparticles, they all move with significant velocities.

However, those velocities are in general not oriented toward the combination of a pair. Rather, the velocity directions are a combination of [a] some component distributed randomly over the particles in essentially all possible directions, and [b] some amount corresponding to a general flow direction.

Table A-1, below summarizes the particle [and antiparticle where applicable] content of interstellar space. The density of the particles, and their related mean distance apart are such as to militate against any significant number of encounters, whether aided by Coulomb attraction or not. [Excepting solar wind, which is local to star's nearby environment, most of the interstellar medium is Hydrogen atoms, not ions.] [Gravitation can be ignored here, it being decades of orders of magnitude weaker than Coulomb attraction.]

Region	Size	Particle Density [/cc]	Particle Energy
Our Solar Wind	Sun Neighborhood	10.	$0.001 - 0.004 \times c$
Our Local Cloud	60 Light Years	0.1	~ 7,000 °K
Our Local Bubble	300 Light Years	0.001	~ 1,000,000 °K
Intergalactic Space	[The Universe]	0.000 ... ?	?

Table A-1 – The Interstellar Medium

As has been pointed out in analyses of our solar wind, with typically *1 atom* in each *10 cm³* of interstellar gas in our local cloud and *10 ions* in each *cm³* of our solar wind, the particles are so far apart that the solar wind and interstellar gas flow through each other without being disturbed by collisions. On that basis, the even less dense regions of the interstellar medium such as ones like our local bubble, those within galaxies in general, and those in intergalactic space are even less conducive to particle / antiparticle encounters.

Another factor bearing on the likelihood of matter / antimatter mutual annihilations occurring in interstellar space is as follows. Because gravitational and Coulomb field attraction communicate at *c*, particles are attracted to where the attractor was, not where it is. That tends to produce orbital motion or "sling shot" non-collision passages rather than direct collisions. For example, a proton traveling at *0.000,001·c [only 300 meters/second]* and at a distance of *0.001 millimeter* from another charged particle [compare that distance with the spacing implied by the densities of the above table] will travel a distance equal to *757 of its proton radii* during the time that its Coulomb field communicates at velocity *c* to the other charged particle its then Coulomb attraction impulse.

All of these various factors taken into account, matter / antimatter collisions must be quite infrequent events in the interstellar medium. When such mutual annihilations occur the appropriate gamma photons are emitted.

Indications of Some Matter / Antimatter Mutual Annihilations

A most likely indication of our detection of cosmic matter / antimatter annihilations is Gamma Ray Bursts [GRB's].

GRB's are flashes of gamma rays coming from seemingly random places in deep space at random times. GRB's last from milliseconds to minutes, and are often followed by "afterglow" emission at longer wavelengths. Gamma-ray bursts are detected by orbiting [*Swift*] satellites about two to three times a week. All known GRB's come from outside our own galaxy. Most GRB's come from billions of light years away [as much as $z = 6.3$ or more].

Under the assumption that a given burst emits energy uniformly in all directions, some of the brightest bursts correspond to a total energy release of 10^{47} *joules*, early a solar mass converted into gamma-radiation in a small amount of time. No candidate process other than a significant matter-antimatter annihilation is able to liberate that much energy so quickly.

Appendix A-1

The Neutron

The fundamental, basic and most simple particles, the proton and the electron and their anti-particles, are developed in the preceding Section 3, *From the Origin to the Complex Universe*. The other usually stable particles, the atomic nuclei and the neutron, are combinations of those basic particles.

The evidence that the neutron is a combination of an electron and a proton is overwhelming.

- Unlike the case with atomic nuclei, where the presence of multiple protons and their mutual electrostatic repulsion makes the nucleus tend to fly apart , an electron and a proton would tend to bind together in mutual electrostatic attraction. No binding energy or mass deficiency would be needed for an electron - proton combination.

- This correlates with the neutron mass, which exceeds the sum of the masses of the hypothesized components, a proton and an electron, by $0.000,839,854$ amu (more than the mass of an electron). The neutron has in this sense a negative mass deficiency or binding energy, a mass excess. One would expect this since the act of combining a proton and an electron should also include at least some of the energy of their mutual attraction.

- Because of the negative binding energy one would expect the neutron to be unstable. While it is stable in a stable atomic nucleus, where it is affected by its overall nuclear environment, free on its own it readily decays into a proton and an electron with a mean lifetime before decay of about 881.5 seconds.

- Of course the combination naturally yields the neutron's electrostatic neutrality.

The primary traditional objection to the concept stems from the matter wave wavelength of the electron. In that view the wavelength associated with the electron component of the proton-electron combination would be far too large and in direct contradiction to observed cross-sections and wavelengths.

However, that objection applies to a "bunch of grapes" concept of the two particles' combination – their, so to speak, sitting side by side like two peas in a pod. But if the two particles combine more intimately into a new neutron form their waves combine more intimately. Figure A-1-1, below, shows the combination of two oscillations at very different frequencies, the higher representing the proton and the lower, the electron of the proton - electron pair of which a neutron would be composed.

95

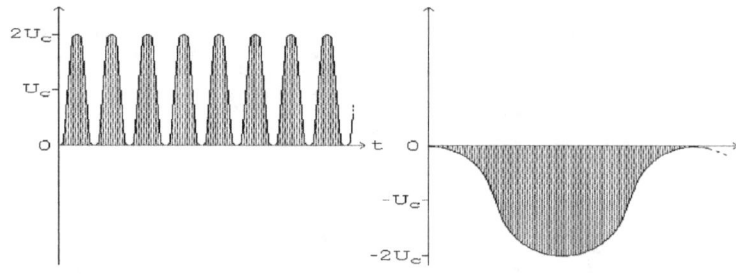

Figure A-1-1(a)
Proton & Electron Oscillations, Two Different Frequencies

As in Figure A-1-1(b), below while our eyes can perceive the longer wavelength in the combined wave form (the envelope), the actual oscillation is only at a wavelength essentially that of the shorter input wavelength. The electron's matter wave need not be a problem.

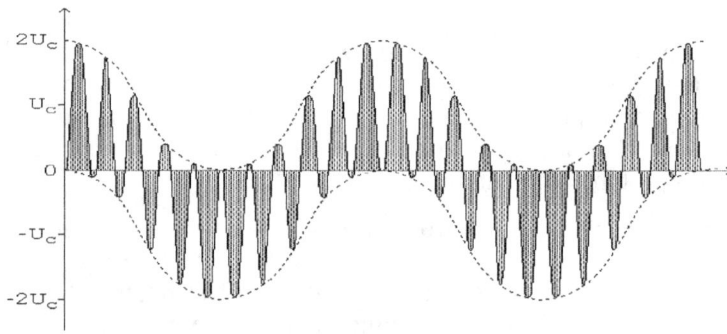

Figure A-1-1(b)
The Sum Oscillation, The Neutron

The *Spherical-Center-of-Oscillation* equation of the neutron, depicted in Figure A-1-1 above, is the sum of those equations of the electron and the proton.

$$(A-1-1) \quad U(_0n^1) = U_c \cdot \left[1 - \text{Cos}(2\pi f_p t) \right] - U_c \cdot \left[1 - \text{Cos}(2\pi f_e t) \right]$$

$$= U_c \cdot \left[\text{Cos}(2\pi f_e t - \text{Cos}(2\pi f_p t) \right]$$

The masses of the proton and electron the combination of which is the neutron are not their rest masses even though their combination in the neutron yields the neutron's rest mass. The component masses are the particles' relativistic masses at high velocity. This comes about as follows.

Since a neutron naturally decays into a proton and an electron those decay particles must be emitted at a velocity equal to or greater than their escape velocity. That is, because the proton and electron strongly mutually attract each other electrically, unless they separate at their mutual escape velocities they will immediately re-combine into a neutron.

Put another way, for a neutron to be formed from a proton and an electron the two must come together from the state of being mutually independent of each other. That means that they must mutually accelerate toward each other. In so doing they will each be at escape velocity and have the kinetic energy of that escape velocity at the moment of their combining into the new particle, the neutron.

The portion of the neutron's overall rest mass that corresponds to the component proton and electron's escape velocity kinetic energy is the neutron rest mass less the sum of the proton and the electron rest masses.

(A-1-2) $\Delta m_n = m_{neutron, \, rest} - [m_{proton, rest} + m_{electron, \, rest}]$

$= 1.008,664,904 - \ldots$

$\ldots - [1.007,276,470 + 0.000,548,579,903]$

$= 0.000,839,854 \text{ amu.}$

In the "classical" sense escape velocity refers to an object of some mass that is gravitationally bound to some other mass, for example a rocket to be launched from Earth. The force attracting the two objects, the rocket and the Earth, to each other acts on them equally in magnitude and opposite in direction. Consequently, momentums that are equal in magnitude and opposite in direction are imparted to them. Since momentum is the product of mass and velocity, when one object (Earth) is much more massive than the other (the rocket) it may be assumed with negligible error that it (the Earth) is not accelerated and its velocity is negligible. Then all of the kinetic energy is attributable solely to the rocket. That kinetic energy must be equal to the gravitational potential energy binding the rocket to the Earth (the two to each other) for the rocket to escape the Earth's gravitational pull.

However, in the case of a proton and an electron the assumption that only the particle of lesser mass is accelerated and that the other particle's kinetic energy is negligible is not valid. It is not that the electron escapes from the proton; they escape from each other. Or, it is not that the electron falls toward the proton; they fall toward each other. The kinetic energy of each is involved and the sum of the kinetic energies must equal or exceed the binding potential energy for the velocities to be at or in excess of escape velocity.

The analysis is as follows (where r is the closest separation between the escaping objects or particles).

(A-1-3)   Gravitational   Electrostatic

Rocket [R] escapes from from Earth [E]	Proton [p] and electron [e] escape each other

(a) PE = Force·r

$$PE = \left[G \cdot \frac{m_R \cdot m_E}{r^2} \right] \cdot r \qquad\qquad PE = \left[\frac{q_p \cdot q_e}{4 \cdot \pi \cdot \varepsilon_0 \cdot r^2} \right] \cdot r$$

(b) Final (escape) Kinetic Energy (KE)
= Initial Potential energy (PE)

$KE_R = PE_{total}$ | $KE_p + KE_e = PE_{total}$

$$\tfrac{1}{2} \cdot m_R \cdot v_R^2 = G \cdot \frac{m_R \cdot m_E}{r}$$

No direct solution

A 2nd relationship is
$|P_p| = |-P_e|$ P is momentum

$$v_{R,esc} = \left[\frac{2 \cdot G \cdot m_E}{r} \right]^{\frac{1}{2}}$$

The two relationships must be simultaneously solved for the velocities

For the gravitational case the escape velocity formulation does not involve the mass of the escaping object. In that sense it is independent of the relativistic mass increase with velocity. Furthermore, in the usual cases treating escape velocity of objects (rocketry and astronautics) the velocity never approaches magnitudes at which significant relativistic effects occur.

However, for the electrostatic case the escape velocity formulation must include the masses of the particles, which masses themselves can vary with their velocity. And, in the case of particles, velocities large enough to involve relativistic effects are likely to occur. Therefore, the electrostatic case must be treated relativistically. The simultaneous solution of the electrostatic case's two equations, kinetic energy and momentum, is as follows.

(A-1-4) <u>Momentum</u>

Magnitude of Proton   Magnitude of Electron
Relativistic Momentum   =   Relativistic Momentum

$$\frac{m_p}{\left[1-\dfrac{v_p^2}{c^2}\right]^{\frac{1}{2}}}\cdot v_p = \frac{m_e}{\left[1-\dfrac{v_e^2}{c^2}\right]^{\frac{1}{2}}}\cdot v_e \qquad m_p \ \& \ m_e \ \text{are rest masses}$$

Solving the above for v_p the following is obtained.

(A-1-5) [<u>Momentum</u> continued]

$$v_p = \frac{m_e \cdot v_e}{m_p\cdot\left[1-\dfrac{v_e^2}{c^2}\right]^{\frac{1}{2}}}\cdot\frac{1}{\left[1+\dfrac{m_e^2 \cdot v_e^2}{c^2\cdot m_p^2\cdot\left[1-\dfrac{v_e^2}{c^2}\right]}\right]^{\frac{1}{2}}}$$

(A-1-6) <u>Energy</u>

Relativistic Energy [As Mass] Is Conserved

$$\left[\frac{KE_p+KE_e}{c^2}\right]_{gained}=\left[\frac{PE_{total}}{c^2}\right]_{lost}$$

$$\left[m_{p,v}-m_{p,rest}\right]+\left[m_{e,v}-m_{e,rest}\right]=m_n-\left[m_{p,rest}+m_{e,rest}\right]\equiv m_{n,\Delta}$$

$$\left[\frac{m_p}{\left[1-\dfrac{v_p^2}{c^2}\right]^{\frac{1}{2}}}-m_p\right]+\left[\frac{m_e}{\left[1-\dfrac{v_e^2}{c^2}\right]^{\frac{1}{2}}}-m_e\right]=m_{n,\Delta}$$

The above equations treat the excess of the neutron's rest mass above the sum of the rest mass of a proton plus that of an electron to be the relativistic KE gained by the two particles in approaching each other from infinite separation distance [per the concept of "escape velocity"].

98

The issue here is: how far apart are the proton and electron in their collision paths toward each other when they have the above kinetic masses, $m_{p,v}$ and $m_{e,v}$? For the calculations to be correct, that is for the hypothesis to be correct, their separation distance at that moment must be such that the two colliding particles are exactly at the moment of combining into the neutron. First the velocities, v_p and v_e, will be calculated by the simultaneous solution of equations *(A-1-4)* and *(A-1-5)*. Then the separation distance of the two particles at the moment of collision will be determined.

(A-1-7) Simultaneous Solution of A-1-4 With A-1-5

The expression for v_p from equation (4) is substituted for v_p in the denominator of the first term of the expression obtained in equation (5). The resulting expression has only v_e unknown and is solved for that value.

Rather than manipulating that expression a computer aided design program is used to calculate selected trial values of v_e until the correct result for $m_{n,\Delta}$ [$m_{n,\Delta} = m_n - m_{p,rest} - m_{e,rest}$] is obtained.

The results of that process are as follows.

(A-1-8) v_e = 275,370,263. m/s

$\qquad\qquad$ = 0.918,536,33 · c

$\qquad v_p$ = 379,350.6975 m/s

$\qquad\qquad$ = 0.001,265,378 · c

At those velocities the proton and the electron have total (relativistic) masses of

(A-1-9)

$$m_{e,total} = \frac{m_{e,rest}}{\left[1-\dfrac{v_e^2}{c^2}\right]^{\frac{1}{2}}} = 2.529,490,15 \cdot m_{e,rest}$$

$$= 0.001,388,308,25 \text{ amu}$$

(A-1-10)

$$m_{p,total} = \frac{m_{p,rest}}{\left[1-\dfrac{v_p^2}{c^2}\right]^{\frac{1}{2}}} = 1.000,000,80 \cdot m_{p,rest}$$

$$= 1.007,276,596 \text{ amu}$$

and their sum is the mass of the neutron.

(A-1-11) $m_{neutron}$ = $m_{p,total}$ + $m_{e,total}$

$\qquad\qquad$ = 1.007,276,596 + 0.001,388,308,25

$\qquad\qquad$ = 1.008,664,904 amu

(These calculations assume that the component proton and electron are in a state of zero momentum and zero kinetic energy before being mutually accelerated into each other. It likewise assumes that the resulting neutron has zero kinetic energy and zero momentum because all the components' kinetic energy goes entirely into the neutron's rest mass and the two component's momentums are equal and opposite in direction netting to zero in combination. To the extent that the components do have initial kinetic energy and momentum then either the resulting neutron will have kinetic energy equal to the sum of the components' initial kinetic energies and momentum equal to the net of the two components' initial momenta or some of those quantities may appear in the form of some type of neutrino given off at the time the particles combine.

(Likewise, in describing the decay of a neutron into a proton and an electron, it was assumed that the neutron initially had zero kinetic energy and zero momentum. To the extent that that is not the case then some form of neutrino will account for the kinetic energy and net momentum not accounted for by the decay product proton and electron.)

THE REMAINING ISSUE IS:
 HOW FAR APART ARE THE PROTON AND ELECTRON IN THEIR COLLISION PATHS TOWARD EACH OTHER WHEN THEY HAVE THE ABOVE KINETIC MASSES ?

Their separation distance at that moment must be such that the two colliding particles are exactly at the moment of combining into the neutron.

An initial calculation of that separation distance, r, is as follows.

(A-1-12)

$$[\text{Potential Energy}_{\text{As Mass}}] \equiv \frac{PE}{c^2} \text{ and must} = m_{n,\Delta}$$

$$\frac{PE}{c^2} = \frac{q_{proton} \cdot q_{electron}}{4\pi \cdot \varepsilon_0 \cdot r} \cdot \frac{1}{c^2} = [0.000,839,854 \text{ amu}] \cdot [\text{kg}/\text{amu}]$$

$$r = \frac{q_{proton} \cdot q_{electron}}{4\pi \cdot \varepsilon_0 \cdot c^2} = \frac{1}{[0.000,839,854 \text{ amu}] \cdot [\text{kg}/\text{amu}]}$$

$$r = 1.840,636,27 \cdot 10^{-15} \text{ meters.}$$

Some years ago experiments involving measurement of the scattering of charged particles by atomic nuclei, yielded an empirical formula for the approximate value of the radius of an atomic nucleus to be

(A-1-13) Radius $= [1.2 \cdot 10^{-15}] \cdot [\text{Atomic Mass Number}]$ meters

which formula would indicate that the proton radius (atomic mass number $A = 1$) is about $1.2 \cdot 10^{-15}$ meters.

The mass of the proton can be expressed as an equivalent energy, $m \cdot c^2$, and that as an equivalent frequency, $m \cdot c^2/h$, or an equivalent wavelength, $h/m \cdot c$. That wavelength (not a "matter wavelength") for the proton is

(A-1-14) $\lambda_p = 1.321,408,96 \cdot 10^{-15}$ meters

quite near to the empirical value for the proton radius from equation *(A-1-11)*.

Thus the initial calculation of the separation distance of the proton and electron when their kinetic masses are just correct for them to form a neutron, equations *(A-1-9)*, *(A-1-10)* and *(A-1-11)*, results in a separation distance of about *1½* proton radii or equivalent wavelengths, equation *(A-1-12)*. That uncorrected result is so close as to essentially validate that the neutron is a combination of a proton and an electron.

However, there is more.

The result at equation *(A-1-12)* must be corrected for a variation in the magnitude of the classical Coulomb interaction as the charges approach near to each other. The direction of the electrostatic effect of a charge is radial to the charge location. At great distances from a charge all of those radii in a local sample are such a small part of the total spherical Coulomb action that they are effectively parallel. But, near to the charge they all effectively diverge (as, of course, they actually do in all cases). That reduces the electrostatic force and requires the charges to approach each other more closely than to the distance calculated at equation *(A-1-12)* – in fact to a separation distance of λ_p exactly, within the limitations of the precision of our data. This develops as follows.

When the two charges are relatively near to each other there is less Coulomb effect because of the radial direction of the Coulomb effect action relative to the charges. Coulomb's law, expressed as potential energy as in equation *(A-1-12)*, above, now becomes as follows.

(A-1-15)

$$[\text{Potential Energy}_{\text{As Mass}}] = \frac{[\text{Reduction Factor}] \cdot \text{PE}}{c^2}$$

$$= [\text{Reduction Factor}] \cdot \frac{q_{\text{proton}} \cdot q_{\text{electron}}}{4\pi \cdot \varepsilon_0 \cdot r} \cdot \frac{1}{c^2}$$

$$\text{and must} = m_{n,\triangle} = [0.000,839,854 \text{ amu}] \cdot [\text{kg}/\text{amu}]$$

But, what is the formulation for the *Reduction Factor*?

For the analysis of the effect of the two charges being so near to each other that the radial divergence of the rays is significant the illustration and dimensions of Figure A-1-2, below, are used. In order to be useful the figure is greatly exaggerated, that is α, β, d and so forth are actually too minute to be seen in an unexaggerated figure.

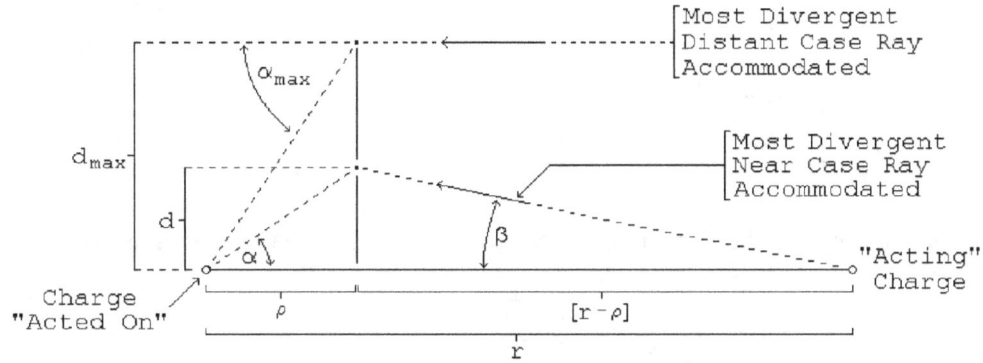

Figure A-1-2
Analysis of Case of Charges Close to Each Other

Even in the case of charges that are far apart, there is only one single ray that is a straight line from one charge to the other. All other rays of electric field must diverge at least minutely from that one straight ray. Therefore, because of the consistent behavior of the Coulomb Law for charges at a variety of separation distances, there is, in effect, a single constant angle of deviation that accommodates those of the divergent rays that enter into the effect. There must be some such angle which is essentially the same for all cases until the charges are close enough that the radial divergence affects the result. That angle is termed α_{max} in this development.

In terms of Figure A-1-2, for the case of charges near to each other, α_{max} must accommodate both β and α. When the charges are far apart β is essentially zero so that $\alpha_{max} = \alpha$. But, the maximum angle, α_{max}, all of which is available to α when the ray source is distant, must, when the ray source is near, account first for removing any ray divergence, β, with any remaining balance left for α. Therefore

$$(A\text{-}1\text{-}16) \qquad \alpha + \beta = \alpha_{max}$$

(The quantity ρ is needed in order for the concept of α_{max} to have meaning; the angle is pointless without defining where it acts. For charges that are far apart α_{max} and ρ are of no significance. When near effects are operating ρ is at $r/2$, half-way between the charges. The concept of ρ is only included here for the initial purpose of presenting in the above Figure A-1-2 the comparison of the near and distant cases.)

The *Reduction Factor* depends upon the reduction of d (of Figure A-1-2) relative to d_{max}, that is the ratio d/d_{max} which quantity is developed as follows.

The angles α, α_{max}, and β are so small that their respective tangents equal their respective angles. Therefore, from the figure

$(A\text{-}1\text{-}17)$

$$\mathrm{Tan}[\alpha_{max}] = \alpha_{max} = \frac{d_{max}}{\rho}$$

$$\mathrm{Tan}[\beta_{max}] = \beta_{max} = \frac{d_{max}}{r - \rho}$$

$$\mathrm{Tan}[\alpha] = \alpha = \frac{d}{\rho}$$

$$\mathrm{Tan}[\beta] = \beta = \frac{d}{r - \rho}$$

From which

$$\alpha = \frac{d}{d_{max}} \cdot \alpha_{max}$$

$$\beta = \frac{d}{d_{max}} \cdot \beta_{max}$$

Then, substituting the above results into equation $(A\text{-}1\text{-}16)$ the following is obtained.

$$(A\text{-}1\text{-}18) \qquad \alpha_{max} = \alpha + \beta$$

$$= \frac{d}{d_{max}} \cdot \alpha_{max} + \frac{d}{d_{max}} \cdot \beta_{max}$$

From which

$$\frac{d}{d_{max}} = \frac{\alpha_{max}}{\alpha_{max} + \beta_{max}}$$

However, α_{max} is a constant quantity (from the consistent Coulomb behavior when the charges are far apart) as is d_{max}.

$$(A\text{-}1\text{-}19) \qquad \alpha_{max} = [\text{A Constant}] \cdot d_{max} \equiv \chi \cdot d_{max}$$

Substituting for α_{max} of equation $(A\text{-}1\text{-}18)$ with equation $(A\text{-}1\text{-}19)$ and for β_{max} of equation $(A\text{-}1\text{-}18)$ with β_{max} of equation $(A\text{-}1\text{-}17)$ the *Reduction*

Factor sought for equation *(A-1-15)* is obtained. It is the d/d_{max} of equation *(A-1-20)*, below.

$$(A-1-20) \qquad \begin{bmatrix} \text{Reduction} \\ \text{Factor} \end{bmatrix} = \frac{d}{d_{max}} = \frac{\chi \cdot d_{max}}{\chi \cdot d_{max} + \dfrac{d_{max}}{r - \rho}}$$

$$= \frac{1}{1 + \dfrac{1}{\chi \cdot [r - \rho]}}$$

This *Reduction Factor* effect is also the cause of the *Lamb Shift*. The Lamb Shift is an extremely slight shifting to higher energy of the inner orbital energy levels of Hydrogen [Coulomb interaction at close separation as analyzed here]. That is, the Lamb Shift is greater as r is smaller. For that reason, it produces a detectable affect principally on the electrons of the inner orbital shells *[n = 1 or n = 2]*.

The form of the effect is depicted graphically in Figure 3, below.

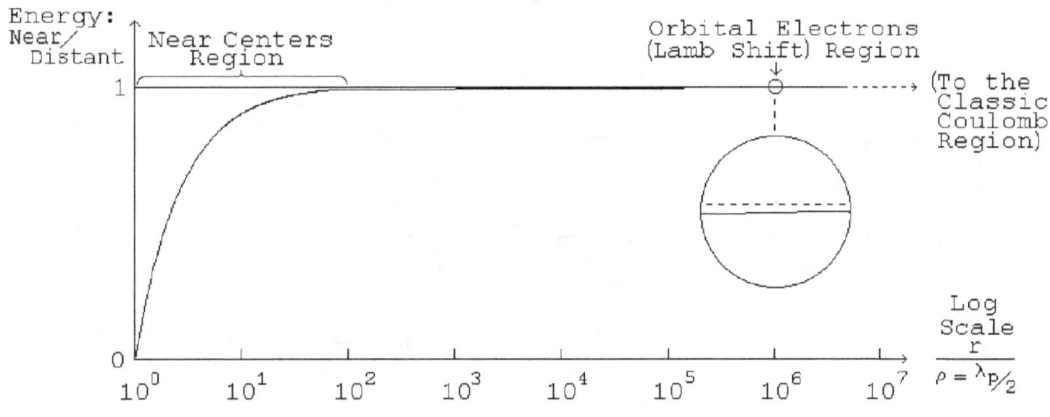

Figure A-1-3
Coulomb Effect <u>Reduction Factor</u> When Charges Are Near to Each Other

The Lamb Shift was attributed to "radiative coupling of the electron to the zero point fluctuation of the vacuum". What that means in plain language is as follows. Heisenberg showed that measurement precision is limited because the information extraction process must change the datum while measuring it. 20th Century physics has questionably extended that to the attribution of a real uncertainty, not merely one of measurement limitation. Then, the zero of the vacuum would also not be precisely zero but a fluctuation in the Heisenberg uncertainty amount about zero. The Lamb shift was attributed to orbital electron interaction with that fluctuation.

The Lamb Shift, is actually caused by the reduction in the negative potential energy due to the orbital electron being near enough to the nucleus that the full Coulomb effect, as when the incoming wave is plane, is slightly reduced as developed above. There being at small values of r marginally less Coulomb attraction, the energy pit in which the electron resides is less deep, which means that its energy is somewhat more than would otherwise be the case. The amount of the effect decreases with increasing r because the reduction in the Coulomb effect decreases as r increases.

The Lamb Shift occurs at much larger values of r (electron orbit radii that are on the order of $r = 10^{-10}\ m)$ than the quite small value of r at which the neutron forms from the combining proton and electron (on the order of $r = 10^{-15}\ m)$. Nevertheless, the Lamb Shift can be used for an approximate calibration of the above *Reduction Factor*. The Lamb Shift is depicted in Figure 4, below.

The shift is stated in terms of the wave number (reciprocal wavelength) because the Rydberg expression for the spectral lines is in terms of wave numbers. The amount of the *Balmer Â* shift is $0.033\ cm^{-1}$. That occurs at the $n = 2$ level where the overall level itself has the term value the Rydberg constant divided by n^2. The fractional shift is then as follows.

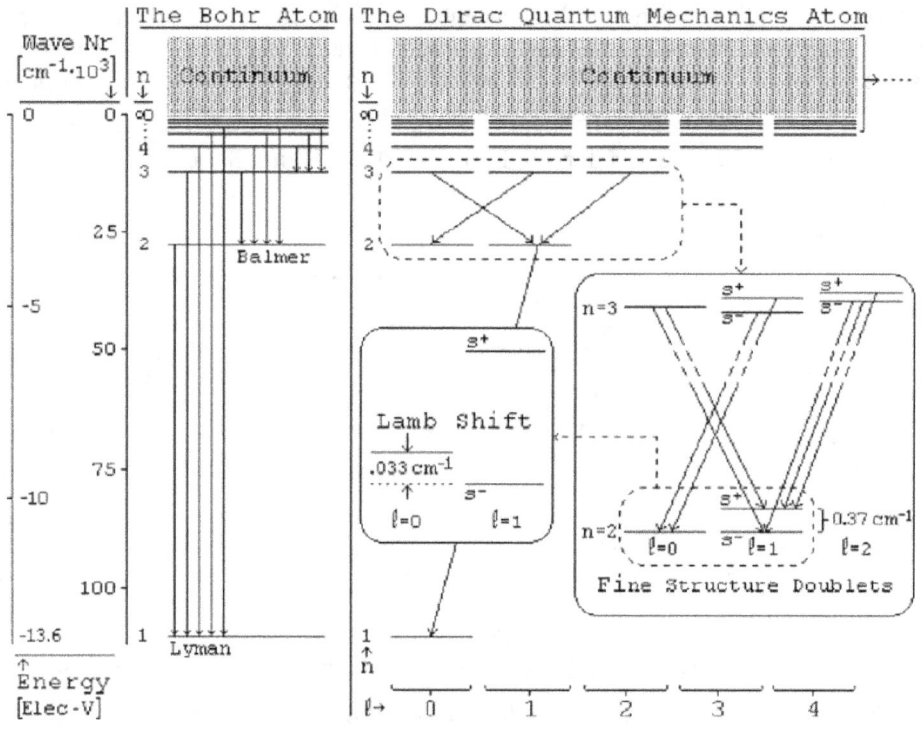

Figure A-1-4
Hydrogen Spectra and the Lamb Shift

$(A-1-21)$ ΔE = Shift = $0.033\ cm^{-1}$ [$n=2$ Balmer $\hat{A}$ shift]

E = Total Wave Number

$$= \frac{Ry}{n^2} = \frac{109,737.31534}{4} = 27,434.3\ cm^{-1}$$

$$\text{Fractional Shift} = \frac{\Delta E}{E} = \frac{0.033}{27,434.3}$$

$$= 1.2 \cdot 10^{-6}\ \text{[dimensionless ratio]}$$

The above *Fractional Shift* is the fractional energy change to the "normal" Coulomb potential energy due to the effect of the two charges being near to each other. The *Reduction Factor* as used in this analysis, equation *(A-1-15)*, is the net energy after that change, *[1 - the above Fractional Shift]* as follows.

104

$$(A\text{-}1\text{-}22) \quad \begin{bmatrix} \text{Reduction} \\ \text{Factor} \end{bmatrix} = [1 - \text{Fractional Shift}]$$

$$= 1 - 1.2 \cdot 10^{-6}$$

$$= 0.999,998,80$$

$$\cfrac{1}{1 + \cfrac{1}{\chi \cdot [r - \rho]}} = 0.999,998,80$$

The radius of the $n = 2$ orbit of Hydrogen is $r = 2.1190152 \cdot 10^{-10}$ m. The ρ in the *Reduction Factor* formula is negligible in the case of the Lamb Shift where $r \approx 10^5 \cdot \rho$ and the precision of the Lamb Shift datum is only two significant digits. Equation *(A-1-20)* can then be solved for the value of χ as follows.

$$(A\text{-}1\text{-}23) \quad \chi = \frac{\text{Reduction Factor}}{r \cdot [1 - \text{Reduction Factor}]}$$

$$= 3.9 \cdot 10^{-15}$$

The general formulation for the *Reduction Factor* is, then, the expression of equation *(18)* with the equation *(22)* value of χ substituted and $\rho = r/2$. The expression for the potential energy as the proton and the electron approach each other to form a neutron is then equation *(A-1-15)* with that *Reduction Factor* substituted. That expression can then be solved for r, the $r_{separation}$ with the following result.

$$(A\text{-}1\text{-}24) \quad r_{separation} = 1.3 \cdot 10^{-15} \text{ meters}$$

The precision of this result is limited to the two significant digits of the Lamb Shift datum. Nevertheless, it is quite close to the wavelength of the proton oscillation in the neutron per equation *(A-1-12)*, $\lambda_p = 1.321,408,96 \cdot 10^{-15}$ *meters*.

Alternatively, if $r_{separation}$ is set at λ_p the resulting value for χ can be calculated and from that the value of ΔE, the Lamb Shift. That calculation gives a Lamb Shift of $.033,611,416$ cm^{-1} compared to the actual datum of $.033$ cm^{-1}.

Two conclusions result from these calculations.

First:
> The cause of the Lamb Shift is the change in the magnitude of the Coulomb effect when the two charges are near to each other not the "radiative coupling of the electron to the zero point fluctuation of the vacuum".

Second:
> The neutron is the combination of a proton and an electron exactly as if each brings to the union its mass equivalent of its escape velocity kinetic energy from the other, the boundary at which the two combine being the wavelength of the proton oscillation, the resulting neutron oscillation being as Figure A-1-1(b) and equation A-1-1 with f_p and f_e being the frequency equivalents of the masses of equations *A-1-10* and *A-1-9* respectively.

Appendix B

The Limitation of the Original Envelopes

This is to show how the otherwise infinite string of envelopes to the original oscillation at the start of the universe was subject to a finite limitation. By "finite limitation" is meant that in the vicinity of the cut-off number of envelopes, N_0, the amplitude of each of the further successive envelopes being imposed on the original $U(t)$, equation 2-5 was successively significantly less than its immediate predecessor and the rate of that amplitude decrease increased sharply with further envelopes – there was a sharp cut-off of amplitude.

After a moderate number of such cut-off region envelopes the amplitude of any further envelopes becomes infinitesimal. While such infinitesimal (and still continuing to become ever more infinitesimal) envelopes theoretically go on to an infinite number of them, the result is equivalent to the convergence to a finite value of a mathematical infinite series such as, for example that of the cosine. The envelopes cut-off is a result of the mathematics of $U(t)$.

The key to that behavior is to be found in Table B-1, below, the expansion of the $Cos^n(x)$ function. The "Cosmic Egg" expression, equation 2-5, repeated below

(2-5) $$U(t) = \pm U_0 \cdot \left[1 - Cos\left[2 \cdot \pi \cdot f_{env} \cdot t\right]\right]^{N_0} \cdot \left[1 - Cos\left[2 \cdot \pi \cdot f_{wve} \cdot t\right]\right]$$

contains the factor

(B-1) $$Cos^{N_0}\left[2\pi(f_{env})t\right]$$

which creates the set of envelopes to the original oscillation. The expansion of the cosine raised to the power of its N_0 exponent behaves according to the pattern illustrated in Table B-1, below. Analysis of the patterns in the coefficients of the individual terms of the $Cos^n(x)$ expansion discloses a pattern related to the binomial expansion as demonstrated in the table.

107

(a) Binomial Expansion Coefficients [a + b]^n

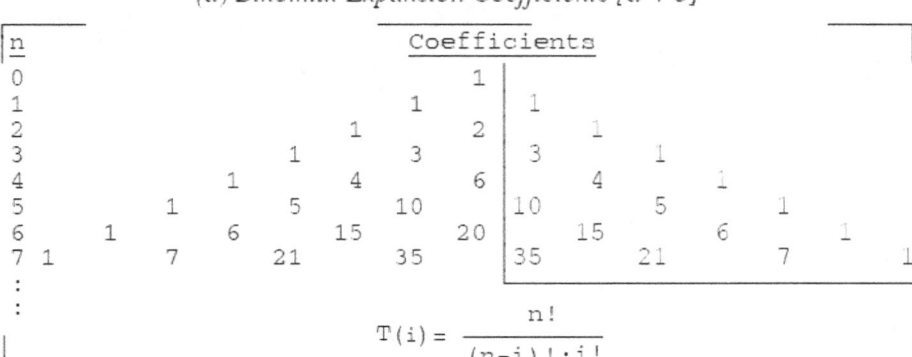

n	Coefficients
0	1
1	1 1
2	1 2 1
3	1 3 3 1
4	1 4 6 4 1
5	1 5 10 10 5 1
6	1 6 15 20 15 6 1
7	1 7 21 35 35 21 7 1
⋮	

$$T(i) = \frac{n!}{(n-i)! \cdot i!}$$

(b) Cos^n(x) Expansion Coefficients

n	Times Cos(*), * =	0x	1x	2x	3x	4x	5x	6x	7x
0		1							
1		–	1						
2		1	–	1					
3		–	3	–	1				
4		3	–	4	–	1			
5		–	10	–	5	–	1		
6		10	–	15	–	6	–	1	
7		–	35	–	21	–	7	–	1
⋮									

$$T(i) = \frac{n!}{(n-i)! \cdot i!}$$

Table B-1

Clearly, with the exception of the constant term (where, in the table, $* = 0x$) the other terms of the expansion of $Cos^n(x)$ have the same coefficients as the corresponding terms of the binomial expansion. The formula for the binomial expansion can thus be used to obtain the coefficients for any value of n in the expansion of $Cos^n(x)$. in the present case for any value of N_0 in the expansion of the $U(t)$ factor

$$Cos^{N_0}\left[2\pi(f_{env})t\right]$$

The cut-off occurs around the value of N_0 regardless of what that value is. Therefore the value of N_0 is not important. Nevertheless it is of interest that various attempts to estimate it give values around 10^{85}.

$N_0 = 10^{85}$ is the n of the formula. It is not practicable and most likely not possible to calculate all of the coefficients of the cosine expansion of the envelopes for 10^{85} envelopes. On the other hand, it is not unreasonable to calculate the 85 cases corresponding to the frequency multiples of the expansion: 10^1, 10^2, 10^3, $\cdots$ 10^{85}.

Figure B-1, below, is a plot of the relative magnitude of the successive coefficients of the various frequency multiples $(1 \cdot x, 3 \cdot x, \cdots 10^{85} \cdot x)$, in the expansion of $Cos^n(x)$ for $n = N_0 = 10^{85}$. The plot indicates a sharp cut-off, an attenuation of the higher frequencies. Figure B-1(a) uses a linear horizontal axis and shows the cut-off in detail. Figure B-1(b) uses a logarithmic horizontal scale to better present the tremendous range in frequency multiples from 1 to 10^{85}. It shows that the cut-off is quite sharp and drastic.

This cut-off is merely the action of the mathematics of $cos^n(x)$.

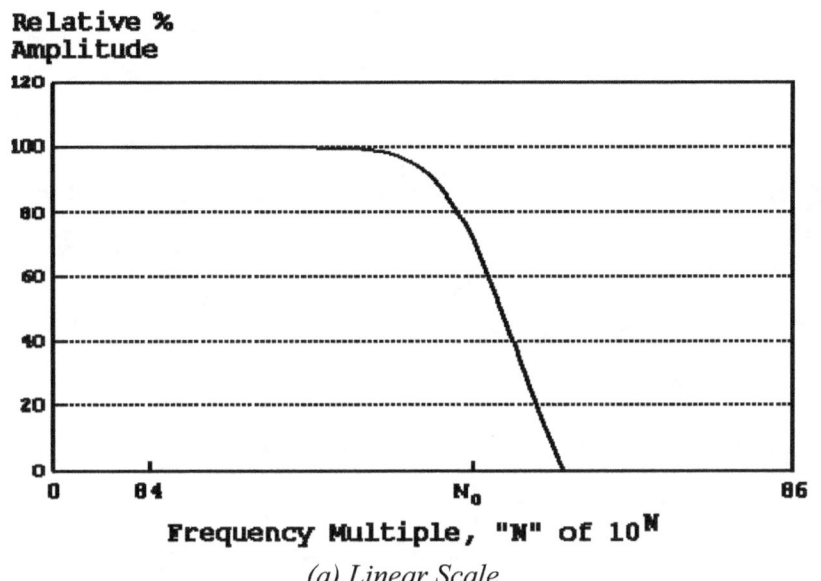

(a) Linear Scale

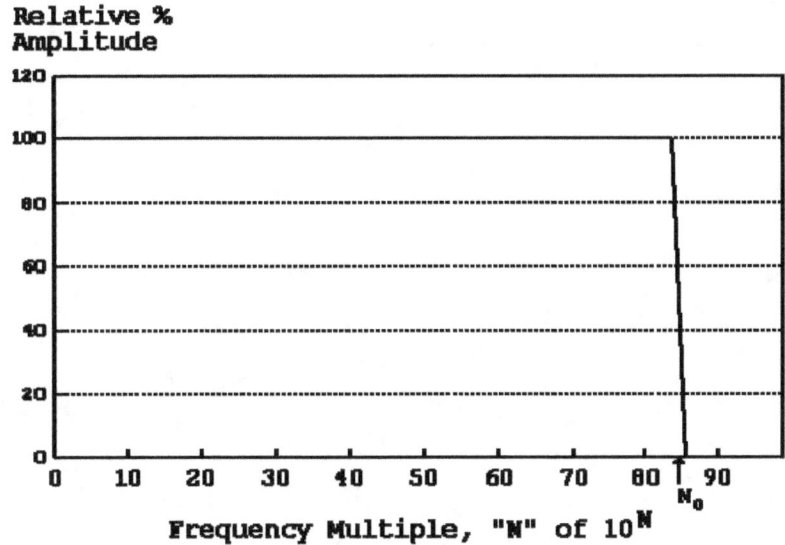

(b) Logarithmic Scale

Figure B-1
The Cos^n(x) Limitation of the "Cosmic Egg

109

Relative Propagated Outward Flow Concentrations: Earth Surface Objects vs. Earth Gravitational Field

Note: In the earlier treatment of the concentrations problem in the book *Gravitics: The Physics of the Behavior and Control of Gravitation* μ_0 and ε_0 were erroneously treated as vector quantities whereas they are actually scalar and so treated here.

FLOW CONCENTRATIONS

For the present purposes the interest is in the potential for slowing of the gravitational *Propagated Outward Flow* Flowing radially outward from the Earth by some configuration of matter at the Earth's surface. The relative amount of slowing depends on the relative amounts or concentrations of the source [Earth gravitation] and encountered [configuration of matter at the Earth's surface] *Propagated Outward Flow* streams.

The problem is, then, to determine within a specified type of matter at the Earth's surface the relative magnitude, u_2, of its ambient *Propagated Outward Flow* as envisioned in Figure 6-4 as compared to u_1, the gravitational Flow propagation arriving from Earth below as envisioned in Figure 6-5. Then the slowing of u_1 by u_2 can be determined.

The Ambient Flow

The ambient Flow within any type of matter is spherically outward from its sources, the atomic components of the matter. Any such stage of this spherical propagation pattern can be split into two hemispheres. That splitting can be chosen to be such that one hemisphere directly faces horizontally. Then, the radially outward rays of that hemisphere all have a component, u_2, in the horizontal direction. That situation is depicted in Figure C-1, below.

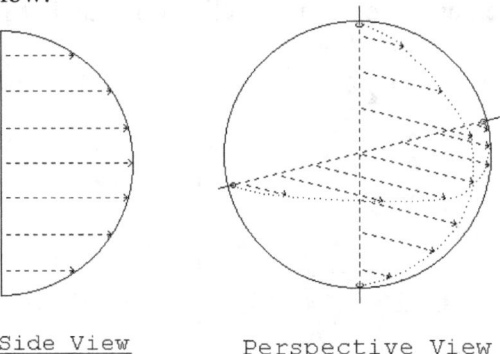

Side View   Perspective View

Figure C-1
Example Rays of u_2 The Horizontal Components of the Radial Outward Flow

Of course, the rays are not discrete rays neatly arranged along a vertical and a horizontal axis. Rather those shown in the figure represent the continuum of medium Flow. All of the rays of the components of u_2 would completely fill the hemisphere volume. The average magnitude of the components corresponds to that hemi-volume divided by the area of the circular base of the hemisphere.

(C-1) r is the radius of the hemisphere, which here
corresponds to the medium amplitude, *u(d)*,
where *d = r*, for a purely radial ray.

$$\text{Volume of Hemisphere} = \frac{1}{2} \cdot \frac{4}{3} \cdot \pi \cdot r^3$$

$$\text{Area of Hemisphere Base} = \pi \cdot r^2$$

$$\text{Average } u_2 = \frac{2}{3} \cdot r \text{ and corresponds to } \frac{2}{3} \cdot [u(d=r)]$$

Some example successive stages of the spherically outward Flow from a single particle Flow source are depicted in Figure C-2, below.

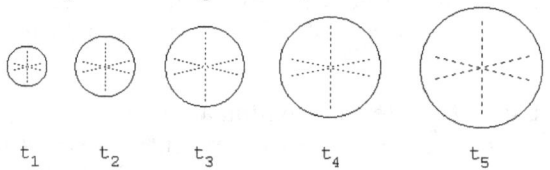

$$t_1 \qquad t_2 \qquad t_3 \qquad t_4 \qquad t_5$$

Figure C-2
Some Stages in a Particle's Spherical Propagation

A single stage, such as that of Figure C-1, of the smoothly continuous sequence of stages of which Figure C-2 is a few intermittent examples, is not a solid hemisphere of medium. Rather it is the wave front of medium propagation at an instant of time. A single stage is the outer surface shell of the hemisphere.

The components of medium Flow pertaining to that shell act at the curved shell surface, not the theoretical flat circular base of the hemisphere of medium Flow. Mathematically one can let the smoothly continuous sequence of such shells be represented by a finite number of nested shells of minute but finite thickness. One such shell is depicted in Figure C-3, below.

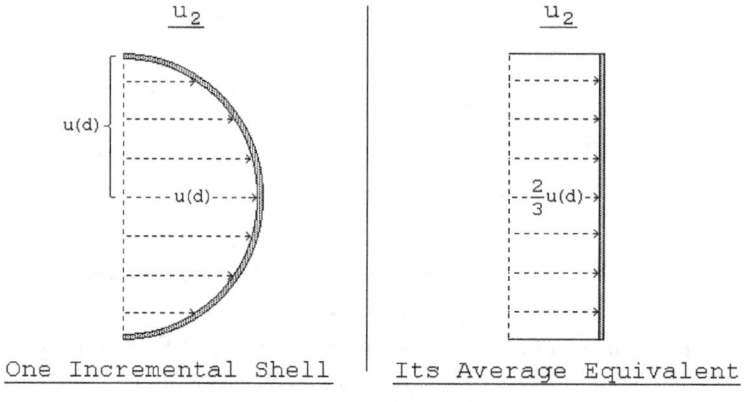

One Incremental Shell | Its Average Equivalent

Figure C-3
A Single Theoretical Shell of Medium Flow

The inverse-square variation of the medium Flow, $u(d)$, with distance, d, from the center of the source particle from which it is propagated is depicted in Figure C-4, below.

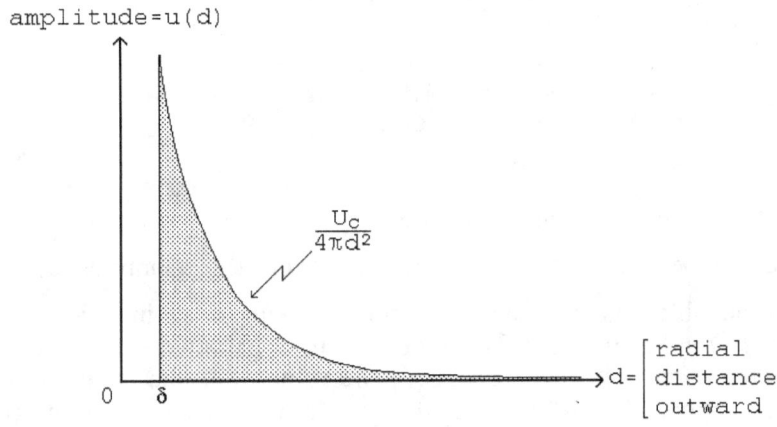

Figure C-4
Flow Amplitude vs. Distance From Center

This amplitude is actually the concentration, the amount of medium per unit area at the surface of a sphere centered on the particle Flow source, as depicted in any single stage of the type depicted in Figure C-2. That amount of medium, itself, is actually the amplitude of the *[1 - Cosine]* form of medium oscillation. [The δ in Figure C-4, above, is the radius of the particle Flow source's core.]

Each atom effectively resides in a cube of side *s*. The particle Flow source of the atom is at the center of the cube and propagates Flow outward in all directions. Per the above Figure C-4, that propagation extends out infinitely in all directions becoming rapidly reduced in magnitude. The cubic volume associated with some single atom experiences the Flow of medium from other adjacent and distant atoms through it in addition to its own propagating medium.

Rather than attempt to sum the myriad varied contributions of all of the other affecting sources in the material to the medium Flow within a particular atom's volume-cube, the same net effect can be obtained by attributing all the action of that particular atom (and each individual atom) as taking place within its own volume-cube. That is, the effect and action per Figure C-4 from $d = \delta$ to ∞ is attributed all to the volume-cube of its source atom with that volume-cube unaffected by medium from other atoms.

Assuming a uniform composition of the matter in question, the matter within which the ambient Flow concentration is to be determined, then the average inter-atomic spacing is the same value as the side of the atom's volume-cube, *s*. That quantity is the cube root of the reciprocal of the density of the matter times the weight of a single component atom.

The maximum hemisphere centered on the center of the atom, the center of the atom's volume-cube, as in Figure C-2, that can fit within the cube of volume allotted to the atom is of radius $R = \frac{1}{2} \cdot S$.

The calculation of *s* is as follows.

(C-2)
$$\text{Density} = \frac{\text{Weight}}{\text{Volume}} = \frac{\text{Atomic Weight}}{s^3}$$

$$s^3 = \frac{1}{\text{Density}} \cdot \text{Atomic Weight}$$

$$= \frac{\text{Total Volume}}{\text{Total Weight}} \cdot \left[\begin{array}{l} \text{Weight of One Atom} = \\ \text{Atomic Mass Number} \times \\ 1.661 \cdot 10^{-27} \text{ kg}/_{\text{amu}} \end{array}\right]$$

$$= \text{Volume for One Atom}$$

$$S = [\text{Volume for One Atom}]^{1/3}$$

Table C-5, below, gives some typical values for these quantities in SI units .

From the table it is clear that inter-atomic spacings, s, in solid elements are on the order of 2.0 to 3.0×10^{-10} meters. In a gas at atmospheric pressure the spacing is on the order of 10^{-9} meters. [The value of δ, the radius of the core of a proton or an electron, is 4.05084×10^{-35} meters, on the order of 10^{-25} times smaller].

Matter	Density	Weight of Atom	Spacing, S
Air	16	25.9×10^{-27}	1.17×10^{-9}
Water	1000	$18. \times 10^{-27}$	2.62×10^{-10}
Carbon	2250	19.95×10^{-27}	2.07×10^{-10}
Aluminum	2700	44.80×10^{-27}	2.55×10^{-10}
Iron	7870	92.88×10^{-27}	2.28×10^{-10}
Lead	11342	345.35×10^{-27}	3.12×10^{-10}

Table C-5
Some Example Inter-Atomic Spacings

The latest medium Flow from the source of u_2, that Flow which has not yet propagated outward and inverse square diffused, has the greatest concentration of medium per area, but it will intercept, interact with, only the smallest target area of external rays, u_1, because it is the smallest shell, analogous to $t1$ of Figure C-2. This is the ray of case "a" in Figure C-6, below.

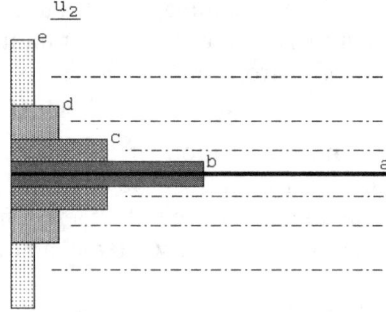

Figure C-6
Encountered Medium Flow for Various Incoming Rays

Medium that had been propagated a moment earlier has progressed somewhat in its inverse square diffusion as in case "b" in Figure C-6. Its concentration of medium per area is less because of the distance that it has propagated, but it will intercept a greater

114

target area of external rays of u_1 for the same reason. The situation is similar but more progressed with the further successive cases in the figure.

An external ray of u_1 that is directed at the center of the encountered particle Flow source will encounter all of the cases depicted in Figure C-6, as indicated in the figure. But an incoming ray that is directed at a point some lateral distance away from the center of the encountered center will encounter only those cases of Figure C-6 which overlay its path.

All of the cases from "a" through "e" and beyond, that is all of the shells from d $= \delta$ to $d = \infty$ can be summed as infinitesimally thick individual shells by integration as follows.

Of the various cases in the above figure, a ray of an intermediate case, such as "c" in the above figure, intercepts all of the cases / shells with a greater radius than the intermediate ray's lateral displacement from the center of the particle Flow source. If we let r represent that lateral displacement of the ray, d the distance outward from the source of u_2 that the shell has traveled, and U_c the fundamental amplitude of the *[1 - cosine]* oscillation, then the summation of the concentrations that that ray encounters in the various shells on outward from lateral displacement r is as follows (the $2/3$ is to deal in average per equation *(C-1)*).

(C-3) $$\frac{2}{3} \cdot \int_{r}^{\infty} \frac{U_c}{4\pi \cdot d^2} \cdot dd$$

This equation *(C-3)* is the product of medium Flow concentration and a distance (the variable of integration, d). That which is needed is the average medium Flow concentration within the atom's volume cube, that is over the range $d = \infty$ to R *(R = ½·s = ½·[the volume cube side])*. The integration on the variable d to ∞ then divided by the distance only out to R attributes all of the atom's medium Flow propagation solely to its own volume-cube.

Therefore, dividing equation *(C-3)* by *[R - δ]* $= R$ because $R \gg \delta$ and performing the integration the equation *(C-4)*, below, is obtained.

(C-4) $$\frac{2}{3 \cdot R} \cdot \int_{r}^{\infty} \frac{U_c}{4\pi \cdot d^2} \cdot dd$$

$$= \frac{U_c}{6\pi \cdot R} \cdot \left[-\frac{1}{d} \right]_{r}^{\infty}$$

$$= \frac{U_c}{6\pi \cdot R \cdot r}$$

In Figure C-6, while a ray of "a" encounters the greatest concentration of medium Flow, only a very minor portion can be in position to experience that concentration. On the other hand, a ray of "c" encounters a reduced medium Flow concentration but a much larger number of rays can have that experience. The number of rays that can experience the medium Flow concentration for any particular lateral displacement, r, is the area of the concentric ring of radius r and thickness dr. For

115

each of the $r's$ of equation *(C-4)* the number of incoming rays of u_1 that encounter that concentration is thus $2\pi \cdot r \cdot dr$.

Therefore, equation *(C-4)*, above, must be integrated by the factor $2\pi \cdot r \cdot dr$ over the range that r can have within the atom's volume-cube, from $r = \delta$ to $r = R$. That process weights each of the different medium Flow concentrations encountered by incoming rays that lie in the successively greater r displacement rings and sums the weighted values. Then dividing that result by the overall target area involved, $\pi \cdot [R^2 - \delta^2] = \pi \cdot R^2$ because $R >> \delta$, gives the average medium Flow concentration contributed by actions within the hemisphere of radius R centered on the center of the source of the *Propagated Outward Flow* and oriented toward the incoming medium Flow.

$$(C-5) \quad \frac{1}{\pi \cdot R^2} \cdot \int_{\delta}^{R} 2\pi \cdot r \cdot [\text{Equation } (A-4)] \cdot dr$$

$$= \frac{1}{\pi \cdot R^2} \cdot \int_{\delta}^{R} 2\pi \cdot r \cdot \frac{U_c}{6\pi \cdot R \cdot r} \cdot dr = \frac{1}{\pi \cdot R^2} \cdot \int_{\delta}^{R} \frac{U_c}{3 \cdot R} \cdot dr$$

$$= \frac{U_c}{3\pi \cdot R^3} \cdot [R - \delta] = \frac{U_c}{3\pi \cdot R^2} \quad [R - \delta = R \text{ because } R >> \delta]$$

This average medium Flow concentration contains the only medium Flow components, u_2, present within the hemisphere within the cube of volume allocated to the atom. That medium concentration must be averaged over the overall cube of atomic volume. The result is the average medium Flow concentration throughout the hypothesized piece of matter.

$$(C-6) \quad \begin{array}{c}\text{Overall}\\ \text{Average}\\ \text{Concentration}\end{array} = \frac{U_c}{3\pi \cdot R^2} \cdot \frac{\text{Hemisphere Volume}}{\text{Atomic Cube Volume}}$$

$$= \frac{U_c}{3\pi \cdot R^2} \cdot \frac{1/2 \cdot [^4/_3 \cdot \pi \cdot R^3]}{S^3}$$

$$= \frac{2 \cdot U_c \cdot [½ \cdot S]}{9 \cdot S^3} \quad [R = ½ \cdot S]$$

$$= \frac{U_c}{9 \cdot S^2}$$

However, this calculation has been for a simple particle Flow source such as a proton or electron. In general, atoms in matter consist of a number of such particles in combination. While their individual Flows are positive and negative, for the purposes of the effect of μ_0 and ε_0 that polarity has no significance; each Flow has the same effect; they do not offset each other.

More precisely the nucleus of an atom is effectively the result of the combining of A protons and $A - Z$ electrons into one overall new particle Flow source oscillating in a complex manner. A is the atomic mass number and Z is the atomic number. The average amplitlude of the complex oscillation of an atomic nucleus is equal to $Z \cdot U_c$.

116

That average value is the result, however, of a $+U$ average value of $A \cdot U_c$ and a $-U$ average value of $[A - Z] \cdot U_c$. That is, the atomic nucleus propagates an average medium amplitude of $A \cdot U_c$ in $+U$ and simultaneously a lesser average medium amplitude of $[A - Z] \cdot U_c$ in $-U$.

Furthermore, the atom's orbital electrons collectively propagate at the same time an average medium amplitude of $Z \cdot U_c$ in $-U$. Those sources of medium Flow are not located at the atomic nucleus, but their average effect is as if they were so located because of their orbits around the atomic nucleus.

The total medium Flow concentration in a piece of solid matter made up solely of atoms of specie $[Z(Element\ Symbol)_A]$ is, then, $A \cdot U_c$ in $+U$ plus $[A - Z] + Z = A \cdot U_c$ in $-U$. That is a collective medium Flow concentration of $2 \cdot A \cdot U_c$. Equation *(C-6)* then becomes as follows for any such matter.

(C-7)   Medium Flow
            Concentration = $\dfrac{2 \cdot A \cdot U_c}{9 \cdot S^2}$
            Within Matter

Using this result, the relative medium Flow concentrations in various forms of matter can be compared. This is done at Table C-7, below, for the same substances as listed in the preceding Table C-5, using the values of $S = [the\ inter-atomic\ spacing]$ from that table.

Matter	Atomic Wt, A	Spacing, S	Ambient Medium
Air	14.99 amu	1.17×10^{-9}	$U_c \cdot 2.43 \times 10^{18}$
Water	18.02 "	2.62×10^{-10}	$U_c \cdot 5.83 \times 10^{19}$
Carbon	12.01 "	2.07×10^{-10}	$U_c \cdot 6.23 \times 10^{19}$
Aluminum	26.98 "	2.55×10^{-10}	$U_c \cdot 9.22 \times 10^{19}$
Iron	55.85 "	2.28×10^{-10}	$U_c \cdot 2.39 \times 10^{20}$
Lead	207.19 "	3.12×10^{-10}	$U_c \cdot 4.73 \times 10^{20}$

Table C-7
Some Example Medium Flow Concentrations, u_2, In Matter

The Incoming Gravitational Flow

Equation *(C-7)* gives the value of u_2, the ambient Flow within matter, which ambient Flow slows the incoming gravitational Flow, u_1. Having determined the value of u_2 it is now necessary to do that for u_1.

The effective gravitational Flow front, u_1, is purely horizontal, that is all rays are vertical, per Huygens Principle applied to the myriad individual wavelets of the myriad gravitating atoms of which the Earth is composed.

The gravitational acceleration produced by one proton acting on a second proton at a separation distance of one meter is as follows.

(C-8)
$$a_g = G \cdot \frac{m_p}{d^2} = (6.67 \cdot 10^{-11}) \cdot \frac{1.67 \cdot 10^{-27}}{1^2}$$
$$= 1.12 \cdot 10^{-37}\ meter/_{second^2}$$

The medium Flow concentration producing that acceleration is as follows.

117

(C-9)
$$u_g = \frac{U_c}{4\pi \cdot 1^2} = U_c \cdot [7.96 \cdot 10^{-2}]$$

The ratio of these two, that is the gravitational acceleration per amount of medium Flow concentration is:

(C-10)
$$\frac{a_g}{u_g} = \frac{1.12 \cdot 10^{-37}}{U_c \cdot [7.96 \cdot 10^{-2}]}$$

$$= \frac{1.41 \cdot 10^{-36}}{U_c} \text{ relative } meter/_{second^2}$$

However, this result is only the case when the source of the gravitational field is a proton having a proton's mass, and, therefore, a proton's Flow oscillation frequency. The gravitational effect is directly proportional to the mass of the source of the gravitational field and the frequency of that source's Flows is directly proportional to its mass.

Therefore, in order to apply in general, equation *(C-10)* must be multiplied by A, the atomic mass in *amu* of the particular gravitational source, divided by *1.07...* the atomic mass in *amu* of a proton.

(C-11)
$$\frac{a_g}{u_g} = \frac{[1.41 \cdot 10^{-36}] \cdot A}{1.07 \cdot U_c}$$

$$= \frac{1.32 \cdot 10^{-36} \cdot A}{U_c} \text{ relative } meter/_{second^2}$$

The ambient Flow concentration in any particular direction in the several substances listed in the preceding Table C-7 then corresponds to the following gravitational accelerations.

Matter	Atomic Wt, A	Ambient Medium	Grav Accel'n
Air	14.99 amu	$U_c \cdot 2.43 \times 10^{18}$	4.81×10^{-17}
Water	18.02 "	$U_c \cdot 5.83 \times 10^{19}$	1.39×10^{-15}
Carbon	12.01 "	$U_c \cdot 6.23 \times 10^{19}$	9.88×10^{-16}
Aluminum	26.98 "	$U_c \cdot 9.22 \times 10^{19}$	3.28×10^{-15}
Iron	55.85 "	$U_c \cdot 2.39 \times 10^{20}$	1.76×10^{-14}
Lead	207.19 "	$U_c \cdot 4.73 \times 10^{20}$	1.29×10^{-13}

Table C-8
Example Ambient Internal Gravitational Accelerations in Matter

For comparison, the value of the Earth's gravitational acceleration at the surface of the Earth is *9.8* $m/_{sec^2}$. Thus the ambient Flow concentrations, as measured by their equivalent gravitational accelerations, available to produce slowing of incoming gravitational Flow of the Earth are on the order of 10^{-15} times too small to have any noticeable effect.

Or, looked at the other way, from equation *(C-11)* the medium Flow concentration corresponding to Earth's gravitational acceleration at the surface is

(C-12)
$$u_g = \frac{U_c \cdot 9.8}{1.32 \cdot 10^{-36} \cdot A} = \frac{7.94 \cdot 10^{36} \cdot U_c}{A}$$

118

The principal components of the Earth are approximately as given in Table C-9, below. From the table the overall average atomic weight, A, of the Earth is about $A = 32.5$.

Earth Component	Percent of Total	Symbol	Atomic Weight	Contribution to Average
Iron	31.0	Fe	55.9	17.3
Oxygen	30.0	O	16.0	4.8
Silicon	16.0	Si	28.1	4.5
Magnesium	15.0	Mg	24.3	3.7
Nickel	2.0	Ni	58.7	1.2
Calcium	1.5	Ca	40.1	0.6
Aluminum	1.3	Al	27.0	0.4
Other	2.0	--	--	--
Earth Average Atomic Weight, A				32.5

Table C-9
Earth Average Atomic Weight, A

CONCLUSION AND RATIOS

Therefore, u_g at the Earths' surface is on the order of

$$u_{gravitational} = u_1 \approx 2 \cdot 10^{35} \cdot u_c$$

using $A = 32.5$ in equation C-12 compared to the ambient Flow concentrations in matter of on the order of

$$u_{local\ ambient} = u_2 \approx 1 \cdot 10^{20} \cdot u_c$$

per the preceding Table C-8 so that

$$u_{gravitational} \approx 10^{15} \cdot u_{local\ ambient}$$

It would thus appear that the medium Flow concentration of Earth surface gravity is so immensely greater than the ambient Flow in local matter that no useful slowing of the Earth's gravitational Flow can be directly effected by a modest amount of matter.

For a useful interaction of matter and gravitational field to take place it would be necessary either to have matter with on the order of 10^{15} times more ambient Flow or a region in space with on the order of 10^{15} times less gravitational Flow or some mixture of those two differences. The former case would require matter of immense density and the latter case gravity so weak that control of it would be of little interest.

Thus the direct use of natural local matter itself to deflect, refract, or otherwise affect or control gravitational Flows appears to be self-defeating in that the amount of matter needed to produce a useful Flow medium concentration would itself be an immense gravitating mass. And, thus, practical control of gravitation requires finding alternative methods of gravitational Flow management.

Such an alternative method is that of Figure 6-7.

THE CAVENDISH EXPERIMENT

In the late 18th Century the British scientist Henry Cavendish measured the gravitational attraction between a pair of <u>lead</u> spheres one weighing 0.73 and the other $158\ kilograms$ separated by a distance of $230\ millimeters$. Comparing the gravitational attraction of the spheres to the Earth's gravitational attraction for the larger one it was found that

119

(C-13) $$\frac{\text{The spheres attraction for each other}}{\text{The Earth's attraction for the larger}} = 3.2 \cdot 10^{-13}$$

a value close enough to the earlier above obtained ratio of

(C-14) $$\frac{u_{local\ ambient} = u_2 \approx 1 \cdot 10^{20} \cdot U_c}{u_{gravitational} = u_1 \approx 2 \cdot 10^{35} \cdot U_c} = 5 \cdot 10^{-14}$$

to approximately validate these calculations and their estimates.

The value 10^{15} for this ratio will be used as being somewhat more conservative an estimate.

APPENDIX D

Factors Affecting Cubic Crystal Tilt

If it were possible to set the minute tilt angle so that the minute offset of $3 \cdot 10^{-19}$ *meter* as called for by the <u>Section 4</u> development could be precisely set and maintained, the "fundamental case", such that the first lattice layer offset is that amount and successive multiples of it sequentially are in the successive layers [2nd layer offset is twice the initial layer; 3rd layer offset is thrice the initial layer; etc.], that direct approach would be taken.

However, the setting of such a minute angle and offset, much less doing so sufficiently precisely, is not practical and probably impossible. To operate using a larger and less precise tilt angle, any tilt angle, the same sufficient number of layers overall required for that "fundamental case" must be employed and the tilt must be such that the actual *x-axis offset* and the actual *y-axis offset* are such that, after that "same sufficient number of layers overall", each required atomic position appears somewhere, in some layer, even though not necessarily in "sequential order".

THE EXACT SUBMULTIPLE OF INTERATOMIC SPACING ISSUE

The most obvious condition of tilt angle and offset that would interfere with "each required effect appearing somewhere, in some layer" would be the actual *offsets* being an exact sub-multiple of the actual interatomic spacing.

For example: with a tilt angle tangent of 0.01, a tilt angle of $0.57°$, the layer-to-layer offset would be 0.01 of an inter-atomic spacing. Layer #2 would be offset $0.01 \times (2.7 \cdot 10^{-10}) = 2.7 \cdot 10^{-12}$ *meter* from layer #1, layer #3 the same from layer #2 ..., and layer #101 would be offset a total of $2.7 \cdot 10^{-10}$ *meter*, the actual interatomic spacing, from layer #1.

In that circumstance any further layers would only reproduce the atom locations relative to the vertical *Propagated Outward Flow* flux that the first 100 layers had introduced.

But if the layer-to-layer offset were such that by layer #101 they accumulated an additional $3 \cdot 10^{-19}$ *meter* total offset from layer #1, then the second 100 layers would deliver atoms all spaced that $3 \cdot 10^{-19}$ *meter* beyond the atoms of the corresponding layers of the first 100 .and the third 100 layers would be correspondingly offset from the second, and so on to ultimately delivering an atom in each of $3 \cdot 10^9$ intervals in each $2.7 \cdot 10^{-10}$ *meter* interatomic space horizontally in the crystal.

123

If that "additional $3 \cdot 10^{-19}$ *meter* total offset" were, instead, any integer multiple of that amount [but still much less than the $2.7 \cdot 10^{-10}$ actual interatomic spacing] the same overall result would obtain – the in effect shuffling of the layers of the cubic crystal lattice.

The actual *offsets* not being an exact sub-multiple of the actual interatomic spacing is essentially automatically assured. The inverse, requiring a perfect integral sub-multiple relationship would be essentially impossible in practice, determined as follows.

A rational number is a number that can be expressed as the ratio of two integers. A rational number expressed as a decimal fraction always exhibits a repetition, over and over, of the sequence of digits in its expression, for example: $0.3333333 \ldots = 1/3$ or $0.125125125 = 1/8$. Conversion of a repeating decimal fraction to the ratio of two integers is done as in equation *(D-1)* on the following page.

Any number exhibiting such a repeating sequence is rational. Any number that does not exhibit such a repeating sequence is not rational, cannot be converted per equation *(D-1)*, above, and is therefore irrational.

Consequently, while the number of rational numbers in any interval is finite, the number of irrational numbers in any interval is infinite.

(D-1) [a] The fraction is defined as "F".
a, b, c, … are digits from the set:
0, 1, … 9
F = 0.abcd … abcd … abcd …

[b] Where n = number of digits in F then

$10^n \cdot$ F = abcd ….abcd … abcd … abcd …

[C] Then, using n = 4 as an example:

$10^4 \cdot$ F – F = abcd

$9999 \cdot$ F = abcd

[d] F = abcd$/9999$

Now consider set *N*, a set of *n* integers [*n* finite]: *1, 2, 3, … , n*. That set is finite, has a finite number of members, is countable and enumerable. Now consider set *R*, all rational numbers such that each such number has a member of *N* as its numerator and a member of *N* as its denominator. There are *n* members of *N*. There are n^2 members of *R*. The number of members of *N* and of *R* is finite.

Now consider set *I*, all irrational numbers greater than *zero* and less than *n*. The number of members of set *I* is infinite. Therefore, the random selection of any number in the interval zero to n, has an infinite probability of being irrational and an infinitesimal chance of being rational.

Therefore, for any installed tilt angle and the offset that it produces, the chance that it would be an exact sub-multiple of the actual interatomic spacing is nil.

On the other hand, the chance that some particular achieved tilt angle and offset requires more layers of cubic crystal than the "fundamental case" because of inefficient scheduling of successive positions is a significant consideration.

TEMPERATURE VARIATION

In addition, a number of variable natural effects are much greater than the precise offset of $3 \cdot 10^{-19}$ *meter*. The effects of temperature variation in the Silicon cubic crystal and various random vibrations within it would overwhelm such a minute setting.

Most materials tend to expand with increase in their temperature. The measure of that effect is the Thermal Coefficient of Expansion, α. That coefficient relates to thermal expansion of the material as in equation *(D-2)*.

```
(D-2)  ΔL = α·L·ΔT

       where:
          ΔT    =    change    in    temperature    in    degrees
                     centigrade*
           L = length of a dimension of the material
          ΔL = change in L due to ΔT
           α = thermal coefficient

       For Silicon α = 3·10⁻⁶ per degree Kelvin*
                        at 20° Centigrade
```

For the interatomic spacing of Silicon the effect of temperature change is per equation *(D-3)*, below.

```
(D-3)  ΔL = α·L·ΔT

          = [3·10⁻⁶]×[2.7·10⁻¹⁰]

          = 8·10⁻¹⁶ meters per degree Kelvin
```

---

* The Kelvin temperature scale has its "zero", i.e. begins at, absolute zero, the lowest possible temperature. The Centigrade [Celsius] temperature scale has its "zero" at 273.15° Kelvin, i.e. they are offset by 273.15°. The magnitude of changes in the two scales are identical; a change of 1° Centigrade = a change of 1° Kelvin.

That as compared to the offset of $3 \cdot 10^{-19}$ *meter* to be created by the tilt. The one-degree temperature variation is over *2,600* times the objective offset. Even a $^1/_{1000}$ degree temperature variation is over double the objective offset. For that reason alone, the setting and maintaining of so precise an objective offset is impractical.

The thermal coefficient of Silicon itself varies with temperature. More precisely it ranges from *2.6* to *3.3* *[× 10⁻⁶]* over the temperature range of *20° to 100° C.*

THERMAL VIBRATIONS AND BLACK BODY RADIATION

The Silicon crystal, at more or less room temperature and as most other materials at that energy, continuously radiates heat energy at frequencies in the infrared range. [The wavelength of infrared radiation is in the range of $3 \cdot 10^{-4}$ to $3 \cdot 10^{-7}$ *meters*, its frequency being in the range of 10^{12} to 10^{15} *Hz [cycles per second]*.] The crystal also simultaneously absorbs the same kind of radiation from other objects. Its atoms are continuously oscillating, vibrating. Such radiation of that energy comes from a reduction in some atoms' oscillations and such absorption is to an increase in some atoms' vibrations.

If the crystal's temperature is greater than its surroundings its radiated energy exceeds that absorbed and it cools down toward thermal equilibrium with its surroundings. Conversely, if it is cooler than its surroundings its temperature increases due to its absorbing more energy than it radiates.

The heat energy corresponding to the crystals' temperature exists in the crystal as the vibratory oscillations of its atoms about their neutral [*temperature = 0° Kelvin*] position. In a crystal lattice the atoms are bound to their average positions by the neighboring atoms. The spectrum of lattice vibrations ranges from low frequencies to ones on the order of 10^{13} *Hz*.

The dependency of atoms' vibrations on its neighbors depends on temperature. At room temperature range most of the thermal energy is in the vibrations of highest frequency. Because of the short corresponding wavelength the motion of neighboring atoms is essentially uncorrelated so that the vibrations can be considered as independently vibrating atoms, each moving about its average position in three dimensions.

[At high temperature they are not independent of each other. At higher temperatures, not applicable to the present analysis the adjacent atoms are more interrelated in their motions and result in oscillatory waves in the crystal lattice.]

The thermal expansion with increase in temperature is due to the increased amount of energy [heat] in the crystal, and the consequent increase in the amplitude of the crystal's vibrations. The above calculated change in length in Silicon per degree centigrade is the change in amplitude of the atom's vibration.

The ΔL *per degree K* [= *degree C*] of $8 \cdot 10^{-16}$ *meters per interatomic space* of equation *(D-2)* is about *1 part in* $3 \cdot 10^5$ of the interatomic space. Being the amplitude <u>change</u> that occurs per degree in a range of *100* or more degrees that implies a larger overall amplitude of on the order of *100* or more times that, $8 \cdot 10^{-14}$ *meters*, or about *0.0003* of the interatomic spacing.

The overall net effect of this is that atomic locations and interatomic spacings are continuously shifting and varying in oscillatory fashion. The amplitude of these shifts is only a small fraction of the interatomic spacing of $2.7 \cdot 10^{-10}$ *meters*.

On the other hand the amplitude of those shifts is on the order of over *250,000* times the $3 \cdot 10^{-19}$ *meter* objective distance from an atom that is sought to be achieved   That is, the lattice thermal vibrations cause the atoms to oscillate back and forth about their nominal neutral position a distance on the order of over *250,000* times the $3 \cdot 10^{-19}$ *meter* objective distance from an atom that it is sought to cause all of the Earth's gravitational field to pass in some layer of the crystal.

That $8 \cdot 10^{-14}$ *meters* oscillatory atomic location range covers over *250,000* desirable or suitable objective atomic locations each one of which is effectively randomly sampled or occupied by the thermal vibrations along with all of the others.

126

THE RANDOM DISTRIBUTION SOLUTION TO THE CRYSTAL TILT

The original concept of the cubic crystal deflector sought to so position the crystal by tilting it relative to the cubic structure of the crystal that atoms of the crystal are forced to effectively occur at successive locations equivalent to a very close, dense positioning of the atoms as seen from the point of view of the purely vertically upward direction of the rays of the Earth's gravitational field. Such positioning would insure that all of the gravitational field is forced to pass extremely close to an atom somewhere in the crystal and to accordingly be deflected away from its natural vertically upward path.

However, the atoms of the Silicon cubic crystal lattice are not fixed in location relative to each other but, rather, are continuously oscillating or vibrating about their nominal neutral positions.

- The vibrations are of various random amplitudes in a range of amplitudes depending on the temperature-determined energy of the vibrations.

- The vibrations are at various random frequencies again in a range of frequencies depending on the temperature.

- The motion of the atoms in their vibration spans a range of locations encompassing a large number of atomic positions that would have been sought to be achieved in various different layers somewhere in the crystal under the original concept and plan.

- That range of motion of the atoms is a small fraction of the neutral interatomic spacing in the crystal.

The net effect of this behavior of the atoms is that the original concept is unworkable. The natural effects of temperature variation and lattice vibrations are much greater than the precise offset intended of $3 \cdot 10^{-19}$ meter. The effects would overwhelm such a minute setting, which is probably too minute to accurately set in any case.

The alternative is to accept the random vibratory behavior of the atoms and incorporate it into the overall design.

First, the vibrations of each atom are largely independent of the behavior of the other atoms because the amplitude of each atom's vibrations are such a small fraction, 0.0003, of the interatomic spacing. At any instant of time the totality of the atoms behaving randomly means that, for a sufficient total number of atoms [a sufficiently thick crystal], every sought position of an atom appears somewhere in the crystal. The range of the atom's vibrations can be thought of as a single "super atom" that simultaneously is at all of the $3 \cdot 10^{-19}$ meter intervals in its range.

Second, the issue of the tilt angle and offset that it produces now is that of properly staggering the atomic vibration ranges of the atoms in each layer that same range amount. That is, the tilt objective is now to offset the second layer from the first layer by one atomic vibration range, 0.0003, of the interatomic spacing, $0.0003 \times 2.7 \cdot 10^{-10} = 8 \cdot 10^{-14}$ meters.

With the "unit" atomic vibration range being $8 \cdot 10^{-14}$ meters then the tangent of the tilt angle to schedule that range at an equal offset, $8 \cdot 10^{-14}$ meters, in each successive adjacent layer is per equation *(D-4)*.

127

(D-4)

$$Tan(Tilt) = \frac{Offset}{Vertical\ Layer\ Thickness}$$

$$= \frac{8 \cdot 10^{-14}}{5.4 \cdot 10^{-10}} = 0.00015$$

Tilt = 0.008°

Figure D-1, below illustrates the effect of equation *(D-4)*.

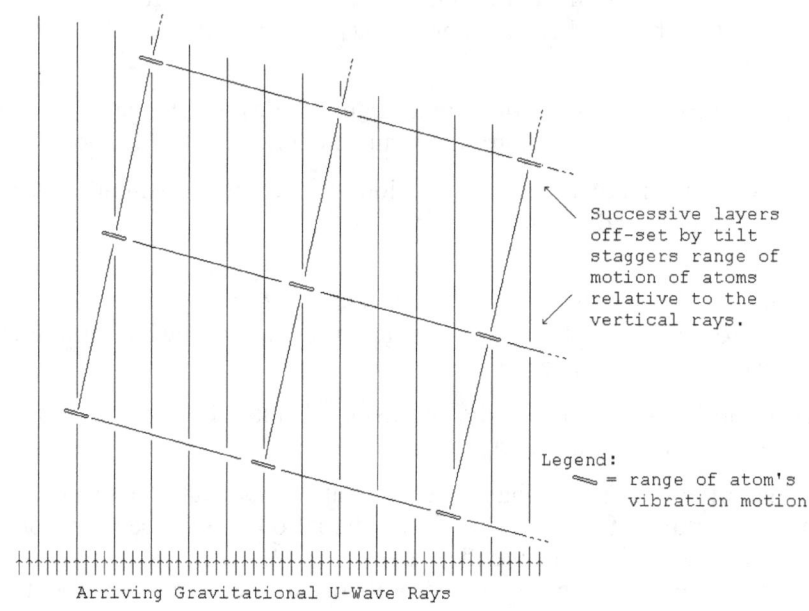

Successive layers
off-set by tilt
staggers range of
motion of atoms
relative to the
vertical rays.

Legend:
~ = range of atom's
vibration motion

Arriving Gravitational U-Wave Rays

Figure D-1
Cubic Crystal Tilt and Atomic Range of Motion Offset
[Not to scale]

If, instead, the layer to layer offset is set at eleven times the "unit" atomic vibration range of $8 \cdot 10^{-14}$ *meters*, that is *[11]* × *[8·10⁻¹⁴] meters* in each successive adjacent layer the tilt is per equation *(D-5)*.

(D-5)

$$Tan(Tilt) = \frac{Offset}{Vertical\ Layer\ Thickness}$$

$$= \frac{[11] \times [8 \cdot 10^{-14}]}{5.4 \cdot 10^{-10}} = 0.002$$

Tilt = 0.1°

That tilt is a not unreasonable value to implement. With it every eleventh layer picks up the position of the second, then third, etc. layer of the equation *(D-4)* case, the layer to layer offset being equal to the atomic vibration range.

128

Using a suitably thick section of a commercially grown Silicon cubic crystal ingot *30 cm* in diameter. the equation *(D-5)* tilt angle tangent of *0.002* would be achieved with a *0.6 mm* thick shim at the edge of the crystal.

Going still farther, if, instead, the layer to layer offset is set at one hundred one times the "unit" atomic vibration range of $8 \cdot 10^{-14}$ *meters*, that is *[101]* × *[8·10⁻¹⁴] meters* in each successive adjacent layer the tilt is per equation *(D-6)*.

(D-6)

$$\text{Tan(Tilt)} = \frac{\text{Offset}}{\text{Vertical Layer Thickness}}$$

$$= \frac{[101] \times [8 \cdot 10^{-14}]}{5.4 \cdot 10^{-10}} = 0.015$$

Tilt = 0.86°

That almost *1°* tilt is a reasonable value to implement. With it every *101st* layer picks up the position of the second, then third, etc. layer of the equation *(D-4)* case, the layer to layer offset there being equal to the atomic vibration range.

Using a suitably thick section of a commercially grown Silicon cubic crystal ingot *30 cm* in diameter. the equation *(D-6)* tilt angle tangent of *0.015* would be achieved with a *4.5 mm* thick shim at the edge of the crystal.

PRECISION AND ERRORS

If the intended *4.5 mm* thick shim were in error by, for example, about *±0.5 mm* then the actual tilt angle tangent would be about *0.013* to *0.017*. That corresponds to the multiple of the "unit" atomic vibration range being about *87.75* to about *114.75*. Either value or any others in that range will eventually produce all of the desired configurations given sufficient layers.

Another precision issue is that of the orientation of the cubic crystal. For the tilt angle to be precise, the bottom of the cubic crystal slab must be exactly one simple layer of the crystal, that is perfectly aligned to the cubic lattice. In addition the surface on which the crystal and shim rest must be perfectly horizontal.

Furthermore, two shims are needed, one for the *x-offset* and one for the *y-offset*. Each must be located at a point on the edge of the crystal corresponding to the midpoint of the interatomic spacing central to the greatest parallel diameter. They must be located at 90° relative to each other, corresponding to two adjacent horizontal sides of the cubic structure.

ANALYSIS OF VARIABLES

There are several quantities or factors bearing on the amount of gravitational deflection produced by a Silicon cubic crystal deflector as here contemplated.

First is the ratio of the *Propagated Outward Flow* concentration of Earth's natural gravitation as compared to the *Propagated Outward Flow* concentration in the light diffracted at a slit, calculated at 10^{15} in <u>Appendix B</u>. This ratio determines the closeness required of the passage of gravitational to atoms of the crystal as calculated in <u>Section 4</u>. This quantity is not variable, but, in spite of <u>Appendix B</u> its value is to some extent an estimate rather than a hard fact. The effect of variation in its value is to correspondingly vary the gravitational *Propagated Outward Flow* passage atomic closeness needed, which in the strict "fundamental case" varies the precise tilt angle called for. Practically, that is of no significance in view of the above "The Random Distribution Solution to the Crystal Tilt".

However variation in the gravitational *Propagated Outward Flow* passage atomic closeness needed produces variation in the number of Silicon cubic crystal layers required which translates into variation in the required thickness of the Silicon cubic crystal slab.

Next is the issue of the vibration range of the crystal's atoms. The range of each of the crystal's atom's vibration varies from that of every other atom because the heat energy so stored varies and continuously changes through exchanges. The atoms' average or typical vibration range is taken above to be about $8 \cdot 10^{-14}$ *meters*. Those ranges and the uniformity over each range of the random distribution of each individual atom's momentary position in the range are approximate and variable.

The extent to which a particular angle of tilt produces comprehensive coverage of the entire crystal by placing the ranges in successive crystal layers exactly adjacent to each other [as viewed by the vertical flowing gravitational *Propagated Outward Flow*s] or, better, sufficiently overlapping, is a variable because the ranges are a variable. The effect of the degree to which that is optimum or not affects the percentage of the total gravitational *Propagated Outward Flow* flux that is deflected.

In addition the atomic vibrations are three-dimensional although the analysis has treated only one-dimensional vibrations as in Figure D-1.

Finally, the calculated required thickness of the Silicon cubic crystal slab or ingot, *49 cm* varies due to all of the above variations. The effect of this is to affect the percentage of the total gravitational *Propagated Outward Flow* flux that is deflected, also. In general, the thicker the slab the more deflection likely to be achieved.

PRELIMINARY DESIGN SUMMARY

Pending the results of further research and development experiments the principle design parameters for the initial cubic crystal gravitic deflector are as summarized below.

Per the calculations of Section 4, a silicon monolithic cubic crystal slab *50 cm* thick or more should result in 100% deflection.

Common commercially produced silicon cubic crystal wafers are on the order of *600 micro-meters [0.6 mm]* thick and up to *30 cm* in diameter. Using commercial wafers of that type with their very small thickness would be impractical.

Therefore a single thick slab is needed such as is commercially produced to form the ingot from which the commercial wafers are sliced.

With regard to the distance from the top of the cubic crystal deflector to the bottom of the object above it, the greater that distance is the more effectively reduced is the gravitational *Propagated Outward Flow* flux acting on the object because the scattered rays of gravitational *Propagated Outward Flow* can more effectively disperse as they have more distance to travel to the vicinity of the object.

The deflector consists of:

- A support having a verified horizontal upper surface for the cubic crystal deflector to rest upon;

- A Silicon cubic crystal slab:

    · *30 cm* in diameter,

    · *50 cm* or more thick, and

    · with the orientation of the cubic structure determined and noted so that the tilt-producing shims can be properly located at the mid-point of two adjacent sides of the horizontal plane of the cubic structure;

- Precision shims *4.5 mm* thick for producing the tilt of the cubic crystal slab: a tilt angle tangent of *0.015* producing a tilt angle of *0.86°.* on the *30 cm* diameter Silicon cubic crystal slab.

Index of Refraction of Propagated Outward Flow vs. of Light

Referring to Figure 6-1 and its related text, the deflection of light is there treated in terms of the indices of diffraction. The development in the present Appendix E is to the effect that index of diffraction is not a factor of significance in the deflection of gravitational field.

The traditional modern physics treatment of the index of refraction has no knowledge of the underlying *Propagated Outward Flow* basis of the light propagation that the index of refraction treats. The traditional index is a composite of the refracting material's affect on the *Propagated Outward Flow* [the slowing affect of direct encounters of the material's *Propagated Outward Flow* and the *Propagated Outward Flow* carrying the light] and the refracting material's electrons' interaction with the light's electromagnetic field.

Based on the various indexes of refraction for various kinds of glass and the index variation with the wavelength of the light passing through them, about 5% or less [depending on the particular material] of the index of refraction, *n*, is the variation of *n vs. frequency*, that due to the light's electromagnetic interaction with the atomic electrons of the material the light passes through.

In Figure E-1, below, *Index of Refraction vs. Wavelength, Frequency*, the variations vs. frequency are parallel and increase in range with glass type approximately in proportion to glass density. The overall levels are also proportional to glass density. The range of the variation vs. frequency is about

$$1.500 - 1.475 = 0.025 \text{ out of } 1.5$$

to about

$$1.800 - 1.720 = 0.080 \text{ out of } 1.8$$

or about 4% to 5% of the total.

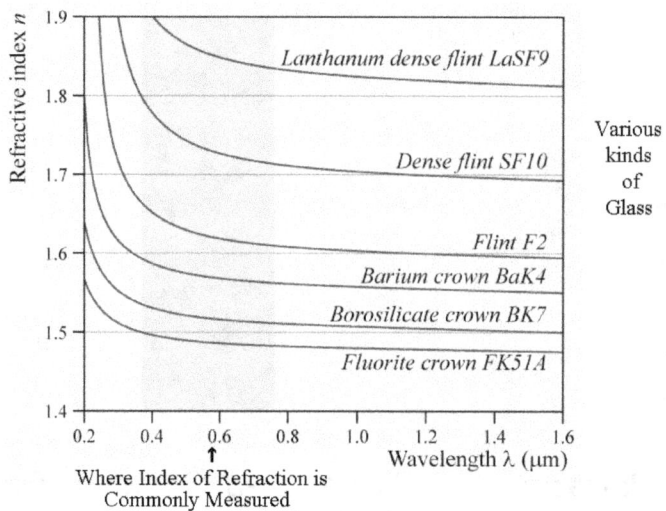

Figure E-1
Index of Refraction vs. Wavelength, Frequency
[Wikipedia, "Index of Refraction"]

The figure would indicate that about the remaining 95% of the index is due to the frequency-independent action of the material on the *Propagated Outward Flow* carrying the light.

However, the index of refraction relationship to *Propagated Outward Flow* propagation depends on the *Propagated Outward Flow* slowing interaction of *Propagated Outward Flow* propagations encountering each other as described in equation *(5-1)* and its related text.

As developed in *Appendix C, Relative Propagated Outward Flow Concentrations: Earth Surface Objects vs. Earth Gravitational Field*, the medium flow concentration of gravitation at the Earth's surface is so immensely greater than the ambient flow in local matter that no useful slowing of the Earth's gravitational flow can be directly effected by a reasonable amount of matter.

Put in other terms, the index of refraction of the Earth's gravitational *Propagated Outward Flow* remains unchanged for practical purposes regardless of the local matter or empty space through which it passes because the ambient *Propagated Outward Flow* concentration of the local matter or empty space through which the Earth's gravitational *Propagated Outward Flow* passes is so minute compared to that of Earth's gravity.

That is, unless some alternative configuration that increases the effectiveness of the ambient *Propagated Outward Flow* concentration in local matter can be found, Finding and developing such a configuration is the case in Section 6.